The First Adventures on Differential Geometry

A Friendly Guide for Beginners

The First Adventures on Differential Geometry

A Friendly Guide for Beginners

Lee Hwee Kuan
A*STAR, Singapore

NEW JERSEY • LONDON • SINGAPORE • BEIJING • SHANGHAI • HONG KONG • TAIPEI • CHENNAI • TOKYO

Published by

World Scientific Publishing Co. Pte. Ltd.
5 Toh Tuck Link, Singapore 596224
USA office: 27 Warren Street, Suite 401-402, Hackensack, NJ 07601
UK office: 57 Shelton Street, Covent Garden, London WC2H 9HE

Library of Congress Control Number: 2024036154

British Library Cataloguing-in-Publication Data
A catalogue record for this book is available from the British Library.

THE FIRST ADVENTURES ON DIFFERENTIAL GEOMETRY
A Friendly Guide for Beginners

ISBN 978-981-12-9617-8 (hardcover)
ISBN 978-981-12-9618-5 (ebook for institutions)
ISBN 978-981-12-9619-2 (ebook for individuals)

For any available supplementary material, please visit
https://www.worldscientific.com/worldscibooks/10.1142/13930#t=suppl

Printed in Singapore

To peace and love for humanity

Preface

Performing mathematical computations on curved surfaces has ubiquitous applications in economics, robotics, graphics, bioinformatics, machine learning, artificial intelligence, chemistry, biology, and clinical sciences. However, the machinery of differential geometry is still not as widely used as it should be, in comparison to its use in theoretical physics. The subject of differential geometry appears too profound for many researchers. Many differential geometry books I have read are designed for students majoring in graduate-level mathematics. Therefore, this subject is unreachable for a vast majority of researchers. Indeed, the subject of differential geometry is not as difficult as many people think. The key unmet need is a lack of learning materials that are designed for beginners. The unique feature of this book is that the materials presented have been brought down to the level of senior high school or freshman mathematics. Prerequisites for reading this book are at the high school or freshman level in:

1. real analysis and multivariate calculus,
2. linear algebra,
3. vector algebra.

This book is not meant to be comprehensive. It touches only on the most essential components of differential geometry to give the reader a good head start. The reader can then build upon these foundations to extend their knowledge by reading other textbooks, which the reader may find unreachable without the foundations this book provides.

To illustrate this point, I have devoted two whole chapters (Chapters 2 and 3) to building up the foundations of geometry on $\mathbb{R}^2$. Performing analysis on $\mathbb{R}^2$ is particularly useful for understanding concepts in higher dimensions because everything can be plotted out and visualized in $\mathbb{R}^2$. Chapter 1 highlights problems in the analysis of surfaces. It also gives an overview of the whole book through oversimplified graphical illustrations using comic strips to tell a story. Chapter 2 focuses on geometries in $\mathbb{R}^2$. It vividly illustrates and explains the concepts of basis and coordinate transformations. Coordinate transformation is a key component of understanding geometry. Chapter 3 goes into analysis, again using $\mathbb{R}^2$ to illustrate concepts. Curvilinear coordinates are introduced to demonstrate coordinate transformations that change points in $\mathbb{R}^2$. This lays the foundations for understanding coordinate transformation in curved spaces. As a concrete example so that readers can grasp the idea and also work through the mathematics, I have used polar coordinates, which are very simple and can be plotted out in two dimensions. The Christoffel symbols are introduced here for polar coordinates.

Only after two chapters (Chapters 2 and 3) of foundation building does the book introduce curved surfaces for the first time in Chapter 4. This chapter is devoted to introducing how to perform calculations on curved surfaces. The coordinate chart is introduced; from coordinate charts, tangent vectors are constructed through infinitesimal changes in the coordinate representations. The geodesic equation is also derived here.

Chapter 5 begins with building differentiation on curved surfaces using covariant derivatives. The covariant derivative is explained in an intuitive and visual way using projections of differentials of tangent vectors onto the tangent plane.

Chapter 6 attempts to introduce concepts that, in my opinion, are the most difficult to grasp. This chapter spends a whole section explaining intrinsic geometry and justifying a new perspective on tangent vectors. Tangent vectors are no longer visualized as an arrow pointing on some surfaces; they are now differential operators. This chapter extensively explains the linkage between these two views of the tangent vector. It also links the mathematical machinery between extrinsic and intrinsic geometry. Covectors, pushforward and pullback are introduced here.

From Chapter 6 onward, all formalism in the book is done in intrinsic geometry notations. Chapter 7 explains flows, their relationship with pushforward and the Lie derivative. This chapter builds foundational material on flows and uses these concepts to explain the Lie derivative and then the Lie bracket. The link between the covariant derivative and the Lie derivative is also demonstrated.

The last chapter of this book, which is Chapter 8, derives the Riemann curvature tensors and their derivatives. It also discusses geodesic deviations and Jacobi equations. From the Jacobi equations, the reader will be able to understand the motivation for deriving the Ricci tensor. Physical interpretations of the Ricci tensor also come from understanding geodesic deviations.

Another feature of this book is that the exercises are designed to guide the reader step-by-step toward understanding the subject. The exercises begin with very simple practices on matrix vector multiplications. All solutions to the exercises are provided next to the questions so that readers can follow and work through them step-by-step. Unlike many mathematics books at the graduate level, this book goes into details of individual mathematical workings without skipping intermediate steps.

In making this book a possibility, I am grateful to the support and feedback from my students, as well as the professional help provided by the various staff of World Scientific Publishing. I also wish to thank Lee Ying Xuan for the cover illustration and Dhanabalan Jeevakaarthik for proofreading my book.

Contents

Chapter 1

Differentials on Surfaces

1.1 Introduction

Mathematical models are ubiquitously used to represent our world. In the early years, mathematics was used in geometry for construction. Calculus was formulated to explain the mechanics of objects. Its applications were extended to many fields in physics, statistics, chemistry, biology, medicine, engineering and computer science. In particular, differentiation and differential equations play a special role in modeling our world in a wide variety of disciplines.

In this section, we discuss briefly and informally functional analysis and calculus. We refer readers to the many, more rigorous mathematics books for a detailed and formal treatment of this subject. Our objective in this chapter is to motivate the need for new mathematical machinery for modeling the physical world with surfaces.

Differentiations or differential operators need space to work on. That means we need to specify the set of continuous, ordinal variables and functions on these variables before we can begin talking about differentials. The most commonly used space is the space of real numbers $\mathbb{R}^n$, where n represents how many numbers we need to represent a point in this space. In order to avoid mixing up these numbers, we assign a symbol that is associated with these numbers. In particular, we assign the symbol e_i that is associated with each of these n numbers. We also specify that the symbols e_i and e_j for $i \neq j$ cannot be added. For example, a point x may be represented as

$$x = x^1 e_1 + x^2 e_2 + \cdots + x^n e_n$$

where $x^i \in \mathbb{R}$, which can be added, subtracted, multiplied and divided.

e_i are objects which cannot be "mixed" by adding and subtracting. They can be multiplied through a special operation called the inner product. A commonly used multiplication operation is

$$e_i \cdot e_j = \delta_{ij} \qquad i, j = 1, \ldots, n,$$

where $\delta_{ij} = 1$ if $i = j$ and equals 0 otherwise. Once we are able to define a point in $\mathbb{R}^n$, we can define a scalar function. Functions are very important objects for modeling our world. For example, every point in our room is represented in $\mathbb{R}^3$, and the temperature at each of the points in our room is given by a function that maps a point in our room to a scalar value in degrees Celsius.

1.2 Euclidean Metric in $\mathbb{R}^n$

Given points in $\mathbb{R}^n$, we would need to define the distance between these points. Using the Euclidean metric, the square length of a vector is defined by

$$\begin{aligned} x \cdot x &= (x^1 e_1 + \cdots + x^n e_n) \cdot (x^1 e_1 + \cdots + x^n e_n) \\ &= \sum_i (x^i)^2 \end{aligned}$$

In the above equation, the Euclidean metric ($e_i \cdot e_j = \delta_{ij}$) is used. The square distance between the two points x and y is given by

$$|x - y|^2 = \sum_i (x^i - y^i)^2$$

1.3 Functions in $\mathbb{R}^n$

We define a function as a mapping from $\mathbb{R}^n$ to $\mathbb{R}$, formally, as

$$f : \mathbb{R}^n \mapsto \mathbb{R}$$

Functions have certain restrictions, such as that they cannot be one-to-many mappings. A precise discussion of functions is a subject that would cover one whole book or at least one chapter of a book. We refer the readers to one of the many functional analysis books for details. To talk about continuity, one has to define limits. Again, we shall not deal with the precise definitions of limits in this book. A function is said to be continuous if it gives the same value as two points it evaluates as they approach each other. The differentiation of a function is, in simple terms, the ratio between the function values at two points x and $x+\epsilon$ and the distance between the two points. For example, for $f : \mathbb{R} \mapsto \mathbb{R}$, the differential is given by

$$\frac{df}{dx} = \lim_{\epsilon \to 0} \frac{f(x+\epsilon) - f(x)}{\epsilon}$$

For functions that map a point in $\mathbb{R}^n$ to $\mathbb{R}$, we define the differentials in each direction:

$$\frac{\partial f}{\partial x^i} = \lim_{\epsilon \to 0} \frac{f(x^1, \cdots, x^i + \epsilon, \ldots x^n) - f(x^1, \ldots, x^n)}{\epsilon} \tag{1.1}$$

1.4 Surfaces

We introduce the concept of surfaces in an intuitive manner. We can think of surfaces generally as subsets of points in $\mathbb{R}^n$. For example, the set of points in our room forms some part of $\mathbb{R}^3$, whereas the surface of the table we are working on forms a more special subset of $\mathbb{R}^3$, such that the surface of the table is a two-dimensional object. Surfaces may not all be flat, like the top of a table; they can be curved, such as the surface of a ball or the surface of a tea cup on the table. Functions on surfaces map every point on the surface to a real number. For example, the temperature of the surface of a tea cup with hot tea inside.

1.4.1 Euclidean geometry in $\mathbb{R}^n$ is insufficient for modeling functions on surfaces

We use $\mathbb{R}^2$ as an illustrative example to highlight the issues with using Euclidean geometry on surfaces. The left panel of Figure 1.1 shows two points A and B on $\mathbb{R}^2$. We can cut $\mathbb{R}^2$ up into equal-size grids and assign horizontal vectors e_1 and vertical vectors e_2 to the grids. e_1 and e_2 are identical at every point. Hence, we can reach point B from point A using translation according to e_1 and e_2 by $3e_1 + 2e_2$. In other words, two unique basis vectors, e_1 and e_2, are enough to reach any point on $\mathbb{R}^2$.

The right panel of Figure 1.1 illustrates the surface of a tea cup. The dimension of this surface is 2, just like the dimension of $\mathbb{R}^2$. However, we can never use two unique basis vectors, e_1 and e_2, to translate between any two arbitrary points. As an illustrative example, we have drawn three non-unique vectors to go from point A to point B on the surface of the tea cup. In summary, the idea of cutting up a surface into an equal-size grid cannot work for surfaces in general. We cannot use a small number of basis vectors to translate between points on a surface. A new mathematical tool is needed to specify points on surfaces.

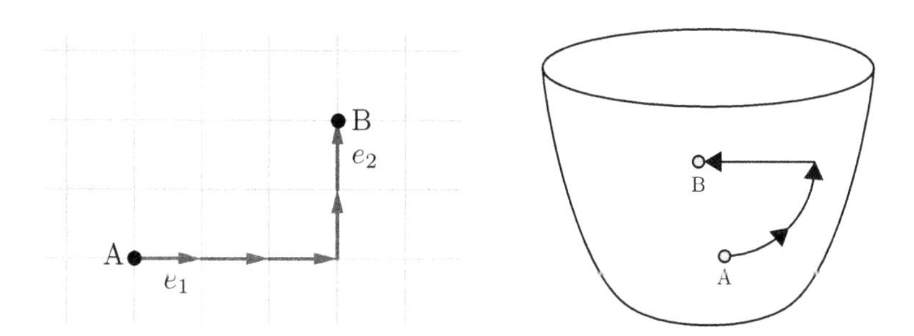

Figure 1.1: The left panel shows that $\mathbb{R}^2$ is cut up into equal-size grids. To reach point B from point A, we perform translation by $3e_1 + 2e_2$. The right panel shows that we cannot cut up the surface of a tea cup using equal-size grids and use two vectors e_1 and e_2 to go from point A to point B.

Functions in Euclidean geometry are defined as mappings from $\mathbb{R}^n$ to $\mathbb{R}$, denoted as $f : \mathbb{R}^n \mapsto \mathbb{R}$. For surfaces, this formalism does not work, and obviously the function $f : \mathbb{R}^n \mapsto \mathbb{R}$ cannot be used on surfaces. In the later chapters of this book, we will learn how to specify points on surfaces. In this case, the function that maps points on surfaces will be modified to take into account how we specify points on surfaces.

1.4.2 Euclidean geometry is insufficient for performing differentiations of functions on surfaces

In $\mathbb{R}^n$, Eq. (1.1) works because both points $(x^1, \ldots, x^i, \ldots, x^n)$ and $(x^1, \ldots, x^i + \epsilon, \ldots, x^n)$ are in $\mathbb{R}^n$, and the function f can be applied to both points. On a surface, however, we cannot specify points in this way or equivalently; points specified in this way do not lie on the surface. Hence, we cannot evaluate functions on points not on the surface. Therefore, Eq. (1.1) becomes invalid on the surface. We cannot differentiate by using the Euclidean geometry approach.

1.4.3 Curvatures and other properties on surfaces

Surfaces possess additional properties not found on $\mathbb{R}^n$. The curvatures of surfaces are an important property that Euclidean geometry would not be able to handle. Another property is the notion of the shortest path connecting two points on a surface. The ideal of the shortest path requires a way to measure infinitesimal lengths on surfaces. Indeed, unlike in Euclidean geometry, where we use identical basis vectors with certain lengths for all points on $\mathbb{R}^n$, the length measure on each point of the surface varies from point to point.

1.5 How to Use This Book

Many textbooks focus on being technically correct and therefore introduce a lot of mathematical formalities. These formalities can be too abstract for the general reader, who is without formal mathematics education. This book approaches the subject with minimal formalities so that it can be more easily digested by the general audience without a rigorous mathematics education. We give up some technical rigor to get more intuition into the subject.

The intention of this book is to provide basic knowledge about how to handle algebraic manipulation on surfaces. This is more of a workbook than a book for reading. The reader will benefit most by working through all exercises and checking the answers provided in the book.

This book should provide enough basic knowledge for the readers to progress into more advanced books in differential geometry. Chapter 2 provides an overview of the mathematics required to work through this book. It also gives a short revision of Euclidean geometry. Chapter 3 introduces functions and fields in $\mathbb{R}^n$. Chapter 4 is where we introduce curved spaces. It introduces tangent planes on surfaces and discusses how to "twist" basis vectors from point to point on the surface. Chapter 5 is devoted to covariant derivatives and parallel transport. The covariant derivative is a central concept in differential geometry. Throughout Chapters 2–5, we work on surfaces embedded in higher dimensions in $\mathbb{R}^n$. From Chapter 6 onward, we work on intrinsic geometry, where the notion of embedding is disregarded. Chapter 6 introduces covectors, pushforward and pullback. Chapter 7 is devoted to flows and the Lie derivative. Chapter 8 introduces the Riemann curvature tensor and its derivatives.

1.5.1 Further readings

This book provides the foundations for further readings in the subject of differential geometry. For a rigorous mathematical treatment of the foundational concepts of differential geometry, the reader can refer to John Lee [8], Manfredo Perdigão Do Carmo [3], Louis Auslander and Robert E. MacKenzie [9] and Yvonne Choquet-Bruhat and Cécile Dewitt-Morette [21]. For intermediate to advanced topics, the following references could be helpful: Manfredo Perdigão Do Carmo [4], Shlomo Sternberg [16], John Lee [7], Dennis Barden and Charles Thomas [2], Bart Vandereycken *et al.* [17], and

Gadea and Muñoz Masqué [13]. Online materials, such as articles and videos, can also be useful references. See, for example, EigenChris [5], Ville Hirvonen [6], and Lee C. Loveridge [10]. For those who are interested in using differential geometry in physics, you can refer to Bernard Schutz [14], Robert M. Wald [18], Carson Blinn [1], Mikio Nakahara [12], and Minser *et al.* [11].

1.6 Adventures of Beng on Differential Geometry

This book introduces many concepts of differential geometry. It can be hard to connect the dots when many concepts are being introduced. The "Adventures of Beng on Differential Geometry" are a set of comic strips that can help summarize different concepts and connect them. It tells the story of Beng's learning journey in differential geometry.

The professor tells Beng that gravity makes space-time curved. Beng exclaims, WHAT?! Space-time seems awfully flat to him. Beng thinks that the professor cannot be lying. So, he searches online for "derivatives of a lie." He found something called Lie derivatives. This set him on a learning journey in differential geometry.

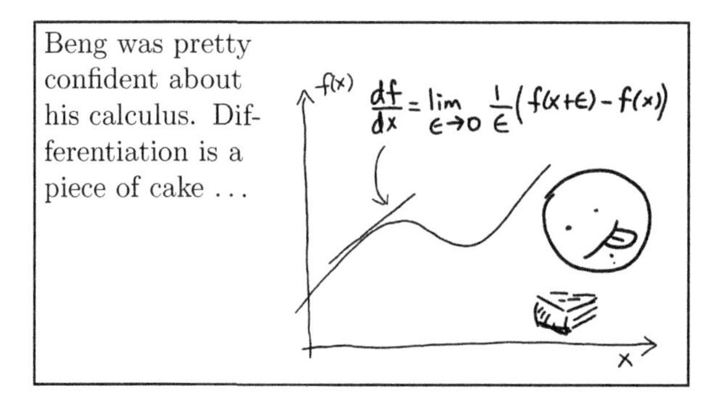

... until he started thinking about how to differentiate on a curved surface, for example, differentiating the temperature at neighboring points on the surface of a tea cup.

Let the surface be a set of points M:

$$x \in M$$

Then, a small displacement Δ from x would bring him to a point outside of M!

$$x + \Delta \notin M$$

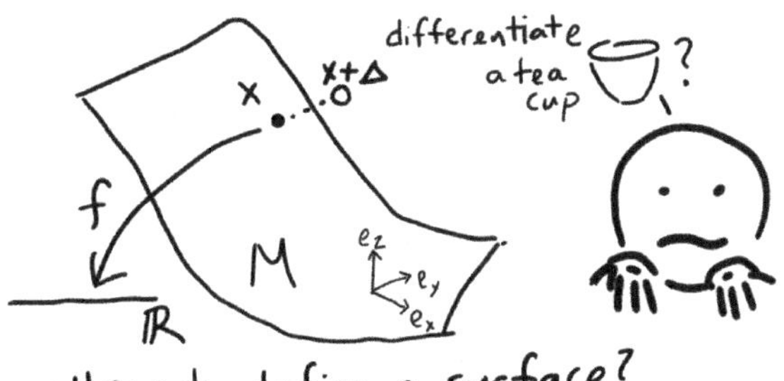

• How to define a surface?

• $x \in M$, $\Delta = \Delta^x e_x + \Delta^y e_y + \Delta^z e_z$

$x + \Delta \notin M$, $f(x+\Delta)$: undefined

He learns that if he parameterized the surface using a chart φ, things got a lot easier using composite functions.

OK

φ x U M

$\varphi(u) \in M \ \forall u$

$f(u) \equiv (f \circ \varphi)(u)$ is always valid

$$\frac{\partial f}{\partial u} = \lim_{\epsilon \to 0} \frac{1}{\epsilon}\left(f(\varphi(u+\epsilon)) - f(\varphi(u))\right)$$

Beng experiments with this new idea and finds out how to construct tangent planes on surfaces by differentiating the chart $\partial\varphi/\partial u^i$.

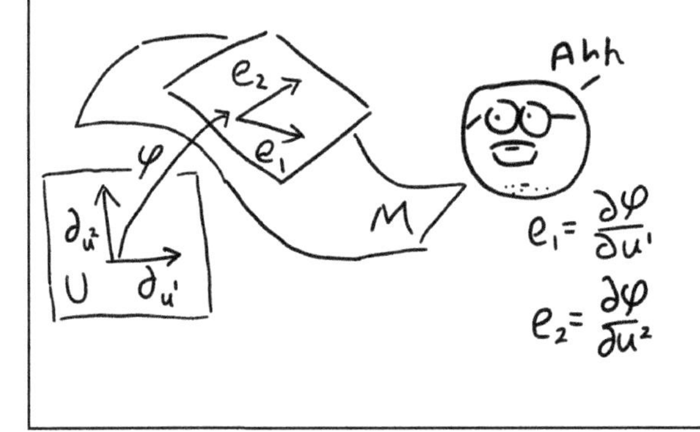

Then, he tries something new. He differentiates the chart twice and notes that the tangent vectors began to twist following the curvature of the surface. He assigns two twisting symbols ${\Gamma^k}_{ij}$ and ${L^k}_{ij}$ to represent this.

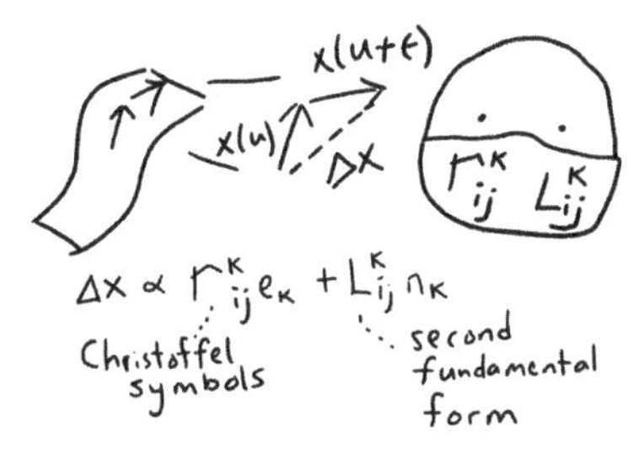

He got so excited with his discoveries that he drew vectors tangent to the surface at every point on the surface. He called his art piece a tangent vector field.

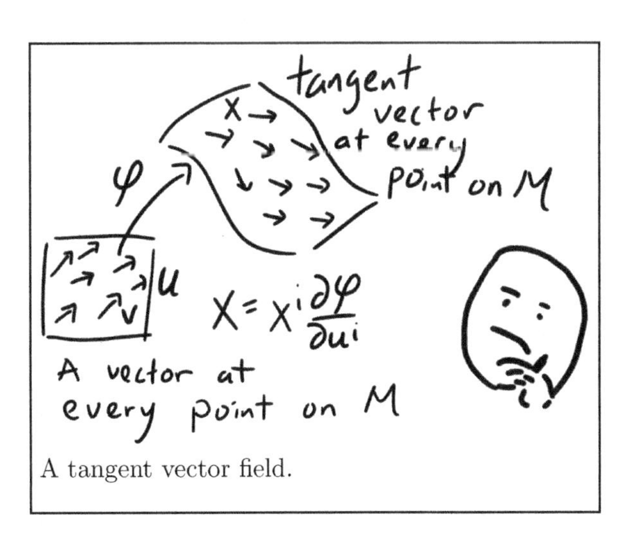

A tangent vector field.

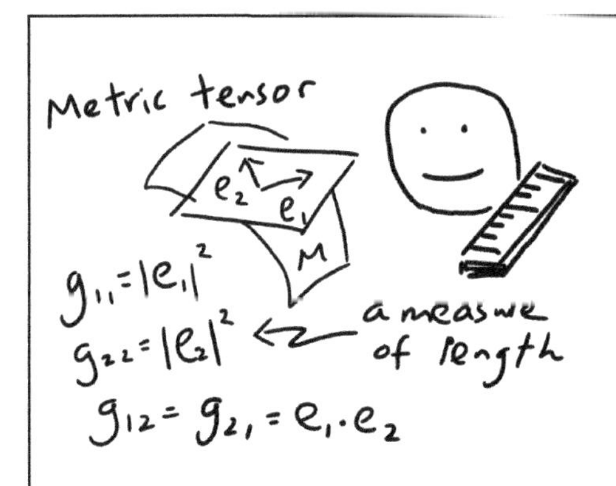

Beng took out his ruler and started to measure the length of the vectors he drew. Since the surface is curved and his ruler is flat, he cannot really do that. He needed a curved ruler that measures length differently at different points of the surface.

Since a flat ruler cannot measure distances on a curved surface, Beng took out a string and began to measure the shortest distance between any two points on the surface.

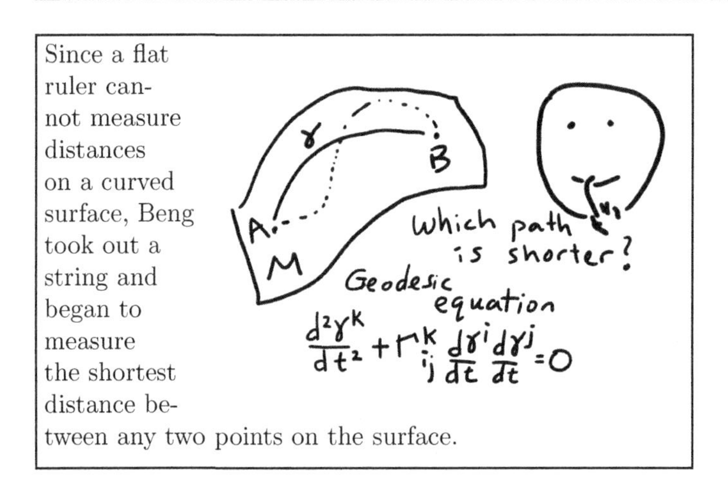

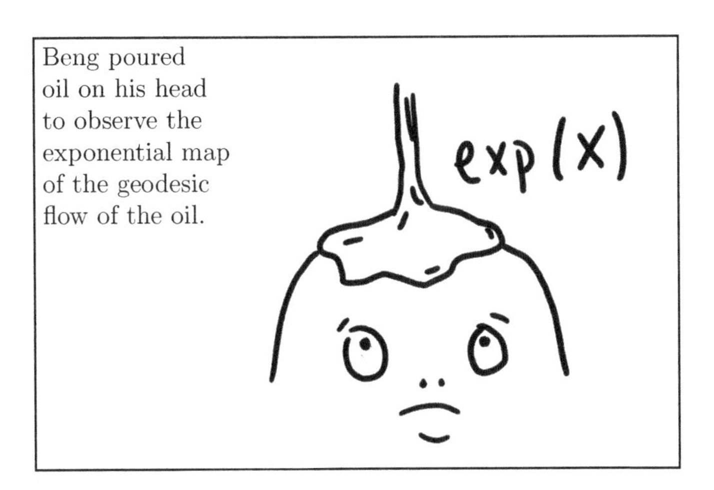

Suddenly, Beng realized something that was not quite right. When he differentiated the vectors, they pointed outward from the surface. This contradicted what he understood about surfaces. For example, the heat flow in the tea cup must be tangential to the surface. Beng projected all vectors pointing out of the surface onto the surface. This is a special kind of derivative, and hence he assigns a special symbol for it: $D_X Y$ is called the covariant derivative, which means to differentiate Y in the direction of X. Also, $D_X Y$ is always tangential to M.

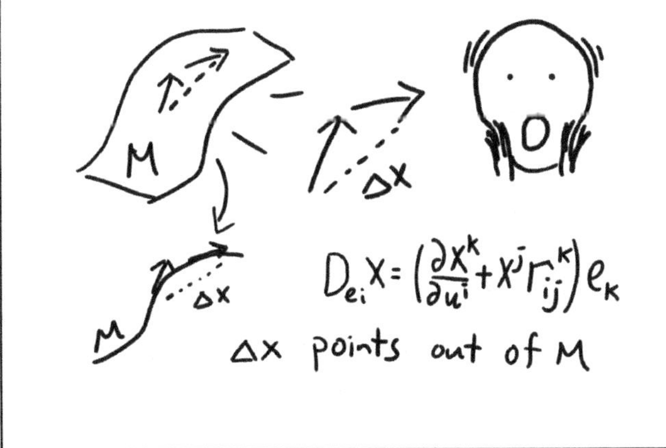

He also started to draw lines on the surface and tried to draw vectors that were parallel along these lines. However, no matter how he drew, those vectors certainly did not look parallel at all!

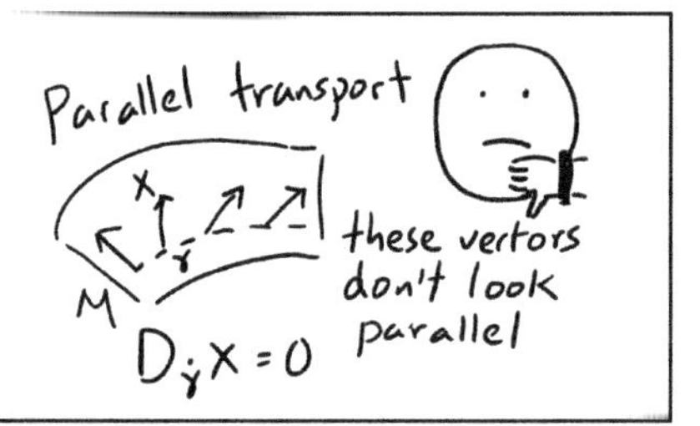

The equations of covariant derivatives only need the coordinate representation u^i and Christoffel symbols, which are derived from the metric tensor alone. The geodesic equation, exponential map and parallel transport all need only the metric tensor. Beng can get rid of the chart φ and work in intrinsic coordinates, in which the tangent vector is no longer a physical vector; it is transformed into a differential operator.

$X \cdots \frac{\partial}{\partial u}$

Beng shared his discoveries with his girlfriend, Lian. She was equally excited. She pointed out to Beng that the vectors looked lonely. So, Beng gave each vector a partner, and he called them vectors and covectors. The relationship between vectors and covectors is as follows. Given a function $f : U \mapsto \mathbb{R}$, a vector takes in f (it differentiates the function f) and gives a number, $X(f) \in \mathbb{R}$. On the other hand, a covector takes in a vector and gives a number, $\omega(X) \in \mathbb{R}$.

Covectors

U

$\omega = \omega_i du^i$

$X = x^i \partial_{u^i}$

$du^i(\partial_{u^j}) = \delta^i_j$

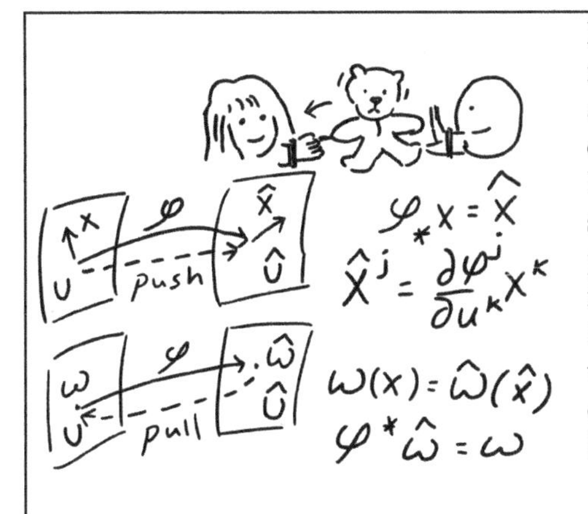

Next, Beng matched vectors and covectors on each point between the two surfaces. Beng pushed vectors from one surface U to another U', while he pulled covectors from U' to U.

Beng also noted he could connect the heads and tails of vectors he drew on the surface and hence trace out flow lines. The lines looked like a flowing river. He recalled the time when he floated down the Kallang River.

He then began to think, what would happen if he took one vector and let it flow along these flow lines? How would this vector twist and turn as it flows along the lines? This is called the Lie derivative, which is given by

$$\mathcal{L}_X Y = [X, Y]$$

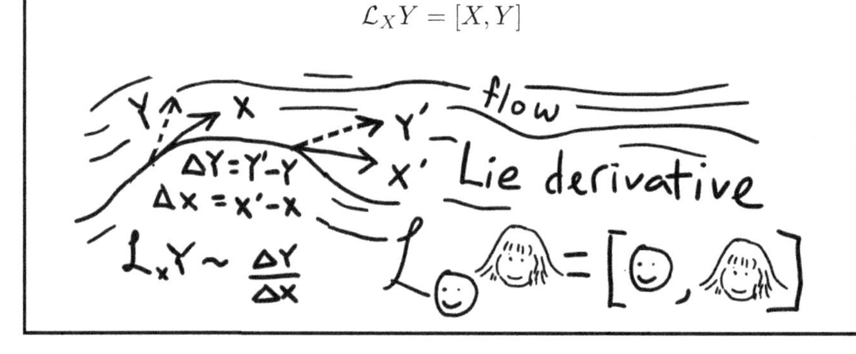

Lian took a pencil and drew parallel vectors on Beng's head to find out how curved the surface of his head was.

Beng, Seng, Lian and Huei set off from the north pole, each flying along the geodesic in different directions. They found that they were flying apart from one another very quickly. However, they will meet again at the south pole.

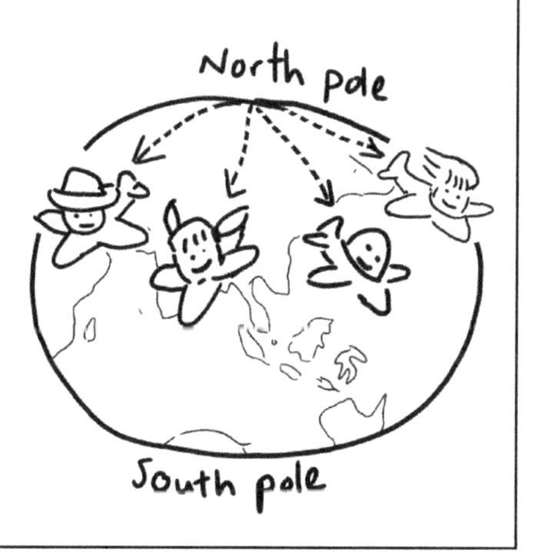

After all these experimentations and exercises, Beng returned to his initial objectives. The professor had said that space-time is curved by gravity. At every moment, Beng must be moving through some curved space-time. Then, he finally understood that if he got close to a black hole, it would be like moving through a tunnel in which his volume would get smaller and smaller following the Ricci curvature of space-time.

Curvatures

$$R(X,Y) = [D_X, D_Y] - D_{[X,Y]}$$

$$D_{\dot{\gamma}} D_{\dot{\gamma}} J + R(J, \dot{\gamma})\dot{\gamma} = 0$$

Black hole

Chapter 2

Mathematical Preliminaries

This chapter lays the foundations for linear algebra and coordinate transformations. The discussions share the context of a two-dimensional Euclidean space. For readers who are not familiar with foundational mathematics in geometry, reading this chapter thoroughly and doing all the exercises will be helpful. For readers who are already familiar with coordinate transformations, spending time to quickly browse through this chapter will help them become familiar with the style and notation followed in this book. Some reference materials related to the subject discussed in this chapter can be found in Philip R. Wallace [19], Frederick Max Stein [15] and William R. Parzynski and Philip W. Zipse [20].

2.1 Coordinate Representation

Consider a set of two abstract object types e_1, e_2. These object types can represent any real-world objects; for example, e_1 can represent apples and e_2 can represent oranges. We define binary operators on these abstract object types, e_1, e_2:

1. Dot product and commutative rule:

$$\begin{aligned} e_1 \cdot e_2 &= g_{12} \in \mathbb{R} \\ e_2 \cdot e_1 &= g_{21} \in \mathbb{R} \\ e_1 \cdot e_1 &= g_{11} \in \mathbb{R} \\ e_2 \cdot e_2 &= g_{22} \in \mathbb{R} \end{aligned}$$

For example, $g_{12} = g_{21} = 0, g_{11} = g_{22} = 1$. When we combine two objects with a dot product, we assign this combination with a real number. This assigned real number, g_{ij}, can have physical meaning. For instance, if we consider apples e_1 and oranges e_2 as very different, we can assign $e_1 \cdot e_2 = 0$. Since apples and apples are the same, we can assign $e_1 \cdot e_1 = 1$.

2. Addition and scalar product: We can do real number algebra on e_1, e_2, such as

$$\begin{aligned} 2e_1 &= e_1 + e_1 \\ 2e_2 + 0.5e_2 &= 2.5e_2 \\ 3e_1 + 4.1e_2 &= 3e_1 + 4.1e_2 \end{aligned}$$

In the last equation, we see that two objects of different types cannot be added.

3. Distributive rule:

$$\begin{aligned} 2(e_1 + 3e_2) &= 2e_1 + 6e_2 \\ (e_1 + 2e_2) \cdot (0.5e_1 - e_2) &= 0.5e_1 \cdot e_1 + e_1 \cdot (-e_2) \\ &\quad + 2e_2 \cdot (0.5e_1) + 2e_2 \cdot (-e_2) \\ &= 0.5g_{11} - 1g_{12} + 1g_{21} - 2g_{22} \end{aligned}$$

We can generalize the above rules to arbitrary n types of abstract objects such that we have $e_1, e_2, \ldots, e_n$. Then,

$$e_i \cdot e_j = g_{ij}$$

g_{ij} in Eq. (2.1) is called the **metric tensor**. The metric tensor will be used frequently throughout this book. We give more meaning to the metric tensor in later sections.

Exercise 2.1.1

1. Given bags of oranges and apples, define the abstract objects e_1, e_2 in terms of the real objects, apples and oranges. Write down the mathematical expression for these bags:

   a. 5 apples and 3 oranges,

   b. 2.5 apples and 1 orange.

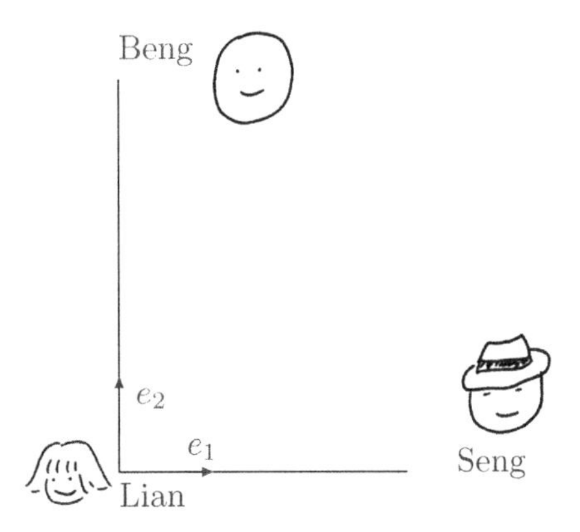

Figure 2.1: Figure for Exercise 2.1.1.

Solution: Let e_1 represent apples and e_2 represent oranges. The bags can be written as

a. $5e_1 + 3e_2$,

b. $2.5e_1 + e_2$.

---

2. Let e_1 represent 0.5 apples and 1.0 orange and e_2 represent 1.0 orange. Write down the mathematical expression for these bags in terms of e_1 and e_2:

   a. 5 apples and 3 oranges,

   b. 2.5 apples and 1 orange.

   Solution: Given that e_1 represents 0.5 apples and 1.0 orange, and e_2 represents one orange,

   a. 5 apples and 3 oranges are given by $10e_1 - 7e_2$.

   b. 2.5 apples and 1 orange are given by $5e_1 - 4e_2$.

---

3. Lian is standing on the peak of a mountain. She sees her friend Beng 4 km to the north of her and another friend Seng 3 km to the east. Figure 2.1 illustrates the geometric relationship between Lian, Beng and Seng.

   a. Write down the mathematical expression of the coordinates of Beng and Seng in terms of e_1 representing 1 km to the east and e_2 representing 1 km to the north.

b. Suppose we let

$$\begin{aligned} e_1 \cdot e_1 &= g_{11} = 1 \\ e_2 \cdot e_2 &= g_{22} = 1 \\ e_1 \cdot e_2 &= g_{12} = 0 \\ e_2 \cdot e_1 &= g_{21} = 0 \end{aligned} \tag{2.1}$$

What is the distance between Beng and Seng?

Solution:

a. The coordinates of Seng and Beng are $3e_1$ and $4e_2$, respectively.
b. The distance between Beng and Seng can be calculated using the Pythagoras theorem. The distance is 5 km.

---

4. Indeed, this example involving Lian, Beng and Seng has a fundamental flaw. What is the flaw?

   Solution: If Lian is standing at the north pole, then "north" and "east" are ill-defined.

2.2 Coordinate Transformation in $\mathbb{R}^2$

In the previous section, we used e_1 and e_2 to represent the objects of interest. There are many alternative symbolic representations. For example, we can make parenthesis with slots like $(\_, \_)$, where in the first slot, we put in the coefficient for e_1, and in the second slot, we put in the coefficient for e_2. We can also use matrix notations in place of parenthesis:

$$x = 0.5e_1 + 2.3e_2 \Leftrightarrow (0.5, 2.3) \Leftrightarrow [0.5, 2.3]$$

The symbol x is used to denote the "position." We use superscripts to represent its components: $x = x^1 e_1 + x^2 e_2$, with $x^1 = 0.5$ and $x^2 = 2.3$ in the above example. Using the matrix notation,

$$\begin{aligned} e_1 &\Leftrightarrow [1, 0] \\ e_2 &\Leftrightarrow [0, 1] \end{aligned}$$

We define the norm of a vector $x = x^1 e_1 + x^2 e_2$ as

$$\begin{aligned} \|x\|^2 = x \cdot x &= (x^1 e_1 + x^2 e_2) \cdot (x^1 e_1 + x^2 e_2) \\ &= (x^1 x^1 e_1 \cdot e_1 + x^1 x^2 e_1 \cdot e_2 + x^1 x^2 e_2 \cdot e_1 + x^2 x^2 e_2 \cdot e_2) \\ &= x^1 x^1 g_{11} + x^1 x^2 (g_{12} + g_{21}) + x^2 x^2 g_{22} \end{aligned} \tag{2.2}$$

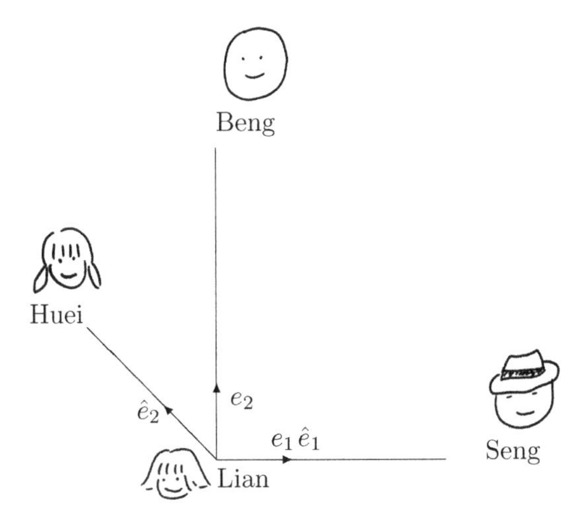

Figure 2.2: Illustration of coordinate transformation in two dimensions.

The above equations can be written in matrix notation as

$$\|x\|^2 = [x^1 \quad x^2] \begin{bmatrix} g_{11} & g_{12} \\ g_{21} & g_{22} \end{bmatrix} \begin{bmatrix} x^1 \\ x^2 \end{bmatrix} \tag{2.3}$$

where we use the standard matrix multiplication rule.

2.2.1 Einstein's double indices summation notation

In geometry, we encounter the summation notation very often; for example, for a point x, we write

$$x = x^1 e_1 + x^2 e_2 = \sum_{i=\{1,2\}} x^i e_i$$

We find that we use the $\sum$ sign a countless number of times. In this book, we apply Einstein's implicit summation notation; that is, we simply leave out the summation sign totally and use double indices to mean "sum over." For example,

$$x = x^1 e_1 + x^2 e_2 = \sum_{i=\{1,2\}} x^i e_i \equiv x^i e_i$$

Referring to the exercise question in the previous section. Lian noticed that Huei is north-west of her. Let $\hat{e}_1$ be 1 km in the easterly direction and $\hat{e}_2$ be 1 km in the north-westerly direction. This is a new coordinate system, which is shown in Figure 2.2. Express e_1 and e_2 given in the previous question in terms of the new coordinate system $\hat{e}_1$ and $\hat{e}_2$ through visual inspection:

$$\begin{aligned}
\hat{e}_1 &= e_1 \\
\hat{e}_2 &= -\frac{1}{\sqrt{2}}e_1 + \frac{1}{\sqrt{2}}e_2 \\
e_1 &= \hat{e}_1 \\
e_2 &= \hat{e}_1 + \sqrt{2}\hat{e}_2
\end{aligned}$$

Writing in matrix form, we have

$$\begin{bmatrix} \hat{e}_1 \\ \hat{e}_2 \end{bmatrix} = \begin{bmatrix} 1 & 0 \\ -\frac{1}{\sqrt{2}} & \frac{1}{\sqrt{2}} \end{bmatrix} \begin{bmatrix} e_1 \\ e_2 \end{bmatrix}$$

$$J^T = \begin{bmatrix} 1 & 0 \\ -\frac{1}{\sqrt{2}} & \frac{1}{\sqrt{2}} \end{bmatrix} \qquad \text{and} \qquad (J^T)^{-1} = \begin{bmatrix} 1 & 0 \\ 1 & \sqrt{2} \end{bmatrix}$$

$$\begin{aligned}
\hat{e}_i &= \sum_k J^k{}_i e_k = J^k{}_i e_k \qquad \text{(Einstein's notation)} \\
e_i &= J^{-1}{}^k{}_i \hat{e}_k
\end{aligned}$$

The transformation matrix J is called the Jacobian. Using the metric tensor in the old coordinate system, e_1, e_2 (Eq. (2.1)), compute the metric tensor $\hat{e}_i \cdot \hat{e}_j$:

$$\begin{aligned}
\hat{g}_{ij} = \hat{e}_i \cdot \hat{e}_j &= J^k{}_i e_k \cdot J^l{}_j e_l \qquad \text{(Einstein's notation)} \\
&= J^k{}_i J^l{}_j e_k \cdot e_l \\
&= J^k{}_i J^l{}_j g_{kl} \\
&= (J^T g)_{il} J^l{}_j \\
&= (J^T g J)_{ij}
\end{aligned}$$

$$\hat{g} = J^T g J \tag{2.4}$$

$$\hat{g} = \begin{bmatrix} 1 & 0 \\ -\frac{1}{\sqrt{2}} & \frac{1}{\sqrt{2}} \end{bmatrix} \begin{bmatrix} 1 & 0 \\ 0 & 1 \end{bmatrix} \begin{bmatrix} 1 & -\frac{1}{\sqrt{2}} \\ 0 & \frac{1}{\sqrt{2}} \end{bmatrix} = \begin{bmatrix} 1 & -\frac{1}{\sqrt{2}} \\ -\frac{1}{\sqrt{2}} & 1 \end{bmatrix}$$

Do the reverse using the metric tensor of $\hat{g}$, and compute the metric tensor of g.

2.2.2 Vectors and components

Consider a vector in the old (e) coordinate basis,

$$x = x^1 e_1 + x^2 e_2 = x^i e_i = [x^1, x^2]$$

What are its components in the $\hat{e}$ coordinate basis?

$$\begin{aligned} \hat{x} &= \hat{x}^k \hat{e}_k \\ &= \hat{x}^k J^i{}_k e_i \\ &= x^i e_i \\ x^i &= \hat{x}^k J^i{}_k \\ \hat{x}^i &= J^{-1}{}^i{}_k x^k \end{aligned}$$

While basis vectors transform from the old to the new coordinate system via forward transformation, vector components transform from the old to the new coordinate system via backward transformation:

$$\begin{aligned} \hat{e}_i &= J^k{}_i e_k \\ \hat{x}^i &= J^{-1}{}^i{}_k x^k \end{aligned}$$

We say that vector components transform in a contravariant way, and basis vectors transform in a covariant way. Referring to Exercise 2.1.1, in the old coordinate system, the position of Beng is

$$x_B = 4e_2 = [0, 4]$$

In the new coordinate system, the position of Beng is

$$\hat{x}_B = \begin{bmatrix} 1 & 1 \\ 0 & \sqrt{2} \end{bmatrix} \begin{bmatrix} 0 \\ 4 \end{bmatrix} = \begin{bmatrix} 4 \\ 4\sqrt{2} \end{bmatrix}$$

Exercise 2.2.1

1. Write the following in Einstein's double indices notation. Whenever possible, express in terms of the metric tensor $g_{ij} = e_i \cdot e_j$:

$$\begin{aligned} a &= x^1 e_1 + x^2 e_2 \\ b &= y^1 e_1 + y^2 e_2 \\ c &= z^1 e_1 + z^2 e_2 \end{aligned}$$

$$a \cdot b$$

$$(a \cdot b)(a \cdot c)$$

$$\hat{e}_i = J^1{}_i e_1 + J^2{}_i e_2$$

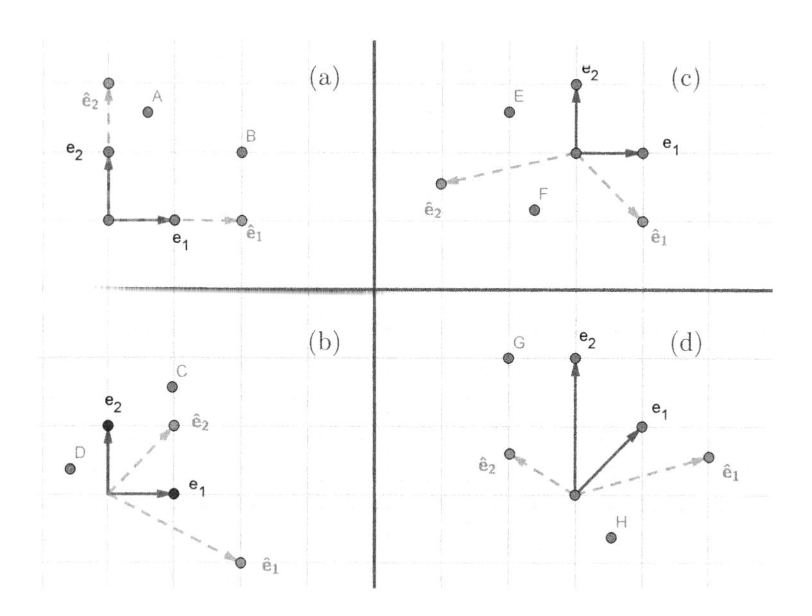

Figure 2.3: Illustration of different coordinate systems.

Solution: Direct matrix multiplication can be used to show the equivalence.

---

2. Let the coordinate transformation for two vectors x, y written in Einstein's double indices summation notation be

$$\begin{aligned} x &= x^i e_i \\ y &= y^i e_i \\ \hat{x} &= \hat{x}^i \hat{e}_i \\ \hat{y} &= \hat{y}^i \hat{e}_i \\ \hat{e}_i &= J^k{}_i e_k \\ \hat{x}^i &= J^{-1}{}^i{}_k x^k \\ \hat{y}^i &= J^{-1}{}^i{}_k y^k \\ J^k{}_i J^{-1}{}^i{}_s &= \delta^k{}_s \end{aligned}$$

Show that

$$x \cdot y = \hat{x} \cdot \hat{y}$$

Solution:

$$\begin{aligned}\hat{x}\cdot\hat{y} &= J^{-1^i}{}_k x^k J^s{}_i e_s \cdot J^{-1^j}{}_t y^t J^l{}_j e_l \\ &= \delta^s{}_k x^k e_s \cdot \delta^l{}_t y^t e_l \\ &= x\cdot y\end{aligned}$$

---

3. Show that Eqs. (2.2) and (2.3) are equivalent by expanding the matrix multiplication of Eq. (2.3)

   Solution: A direct matrix multiplication will show the equivalence of both equations.

---

4. Referring to Exercise 2.1.1, the effort needed for Seng to reach Beng should be invariant to the coordinate system being used. Compute the effort for Seng to reach Beng in the old and new coordinate systems and compare your answers.

   Solution: In the old coordinate system,

$$\Delta x = x_S - x_B = [3, -4]$$

$$\|\Delta x\|^2 = [3,-4]\begin{bmatrix}1 & 0\\ 0 & 1\end{bmatrix}\begin{bmatrix}3\\ -4\end{bmatrix} = 25$$

In the new coordinate system, the positions of Beng and Seng are

$$\hat{x}_B = \begin{bmatrix}4\\ 4\sqrt{2}\end{bmatrix}$$

$$\hat{x}_S = \begin{bmatrix}1 & 1\\ 0 & \sqrt{2}\end{bmatrix}\begin{bmatrix}3\\ 0\end{bmatrix} = \begin{bmatrix}3\\ 0\end{bmatrix}$$

$$\Delta\hat{x} = [1, 4\sqrt{2}]$$

$$\|\Delta\hat{x}\|^2 = [1, 4\sqrt{2}]\begin{bmatrix}1 & -\frac{1}{\sqrt{2}}\\ -\frac{1}{\sqrt{2}} & 1\end{bmatrix}\begin{bmatrix}1\\ 4\sqrt{2}\end{bmatrix} = 25$$

---

5. In Exercise 2.1.1, we use $g_{ij} = \delta_{ij}$. Consider

$$g = \begin{bmatrix}1 & 0\\ 0 & 2\end{bmatrix}$$

Compute the distance between Seng and Beng using the old and new coordinate systems.

Solution: In the old coordinate system,

$$\Delta x = [3, -4]$$

$$\|\Delta x\|^2 = [3, -4] \begin{bmatrix} 1 & 0 \\ 0 & 2 \end{bmatrix} \begin{bmatrix} 3 \\ -4 \end{bmatrix} = 41$$

In the new coordinate system, first compute the metric tensor:

$$\hat{g} = (J^T g J) = \begin{bmatrix} 1 & -\frac{1}{\sqrt{2}} \\ -\frac{1}{\sqrt{2}} & 3/2 \end{bmatrix}$$

Use the metric tensor to compute the distance between Seng and Beng:

$$\|\Delta \hat{x}\|^2 = [1, 4\sqrt{2}] \begin{bmatrix} 1 & -\frac{1}{\sqrt{2}} \\ -\frac{1}{\sqrt{2}} & 3/2 \end{bmatrix} \begin{bmatrix} 1 \\ 4\sqrt{2} \end{bmatrix} = 41$$

---

6. Refer to Figure 2.3. For each figure:

    1. Write down the components of the points, A, B, C, D, E, F, G and H in the e_1, e_2 and $\hat{e}_1, \hat{e}_2$ coordinates.
    2. Compute the Jacobian for transforming from e_1, e_2 to $\hat{e}_1, \hat{e}_2$.
    3. Given $e_1 \cdot e_1 = e_2 \cdot e_2 = 1$ and $e_1 \cdot e_2 = e_2 \cdot e_1 = 0$, compute the metric tensor for $\hat{e}_1, \hat{e}_2$.

Solution: First, we compute the Jacobian and its inverse. Then, the components of vectors transform in a contravariant way by the inverse of the Jacobian. The Jacobian and its inverse for the figures are

for (a) $\quad J = \begin{pmatrix} 2 & 0 \\ 0 & 2 \end{pmatrix} \quad J^{-1} = \begin{pmatrix} 0.5 & 0 \\ 0 & 0.5 \end{pmatrix}$

for (b) $\quad J = \begin{pmatrix} 2 & 1 \\ -1 & 1 \end{pmatrix} \quad J^{-1} = \frac{1}{3} \begin{pmatrix} 1 & -1 \\ 1 & 2 \end{pmatrix}$

for (c) $\quad J = \begin{pmatrix} 1 & -2 \\ -1 & -0.4 \end{pmatrix} \quad J^{-1} = -\frac{1}{2.4} \begin{pmatrix} -0.4 & 2 \\ 1 & 1 \end{pmatrix}$

for (d) $\quad J = \begin{pmatrix} 2 & -1 \\ -0.7 & 0.8 \end{pmatrix} \quad J^{-1} = \frac{1}{0.9} \begin{pmatrix} 0.8 & 1 \\ 0.7 & 2 \end{pmatrix}$

The components of the position vectors are transformed in a contravariant way using J^{-1}:

$$\begin{array}{rr}
A = 0.6e_1 + 1.6e_2 & B = 2.0e_1 + 1.0e_2 \\
A = 0.3\hat{e}_1 + 0.8\hat{e}_2 & B = 1.0\hat{e}_1 + 0.5\hat{e}_2 \\
C = 1.0e_1 + 1.6e_2 & D = -0.6e_1 + 0.4e_2 \\
C = -0.2\hat{e}_1 + 1.4\hat{e}_2 & D = -(1/3)\hat{e}_1 + (0.2/3)\hat{e}_2 \\
E = -1.0e_1 + 0.6e_2 & F = -0.6e_1 - 0.8e_2 \\
E = -0.667\hat{e}_1 + 0.167\hat{e}_2 & F = 0.57\hat{e}_1 + 0.58\hat{e}_2 \\
G = -1.0e_1 + 1.5e_2 & H = 0.6e_1 - 0.6e_2 \\
G = 0.78\hat{e}_1 + 2.56\hat{e}_2 & H = -0.13\hat{e}_1 - 0.87\hat{e}_2
\end{array}$$

Use transformation of the metric tensor:

$$\begin{aligned}
\hat{e}_i \cdot \hat{e}_j &= J^k{}_i e_k \cdot J^m{}_j e_m \\
\hat{g}_{ij} &= J^k{}_i J^m{}_j g_{km} \\
&= J^k{}_i J^m{}_j \delta_{km} \\
&= J^1{}_i J^1{}_j \delta_{11} + J^1{}_i J^2{}_j \delta_{12} + J^2{}_i J^1{}_j \delta_{21} + J^2{}_i J^2{}_j \delta_{11} \\
&= J^1{}_i J^1{}_j + J^2{}_i J^2{}_j
\end{aligned}$$

Write J in matrix form:

$$J = \begin{pmatrix} J^1{}_1 & J^1{}_2 \\ J^2{}_1 & J^2{}_2 \end{pmatrix}$$

for (a) $J^1{}_1 = J^2{}_2 = 2$, $J^1{}_2 = J^2{}_1 = 0$.

$$\begin{aligned}
\hat{e}_1 \cdot \hat{e}_1 &= 4 \\
\hat{e}_2 \cdot \hat{e}_2 &= 4 \\
\hat{e}_1 \cdot \hat{e}_2 &= 0 \\
\hat{e}_2 \cdot \hat{e}_1 &= 0
\end{aligned}$$

for (b) $J^1{}_1 = 2$, $J^2{}_1 = -1$, $J^1{}_2 = 1$, $J^2{}_2 = 1$.

$$\begin{aligned}
\hat{e}_1 \cdot \hat{e}_1 &= J^1{}_1 J^1{}_1 + J^2{}_1 J^2{}_1 = 4 + 1 = 5 \\
\hat{e}_2 \cdot \hat{e}_2 &= J^1{}_2 J^1{}_2 + J^2{}_2 J^2{}_2 = 1 + 1 = 2 \\
\hat{e}_1 \cdot \hat{e}_2 &= J^1{}_1 J^1{}_2 + J^2{}_1 J^2{}_2 = 2 - 1 = 1 \\
\hat{e}_2 \cdot \hat{e}_1 &= J^1{}_2 J^1{}_1 + J^2{}_2 J^2{}_1 = 2 - 1 = 1
\end{aligned}$$

for (c) $J^1{}_1 = 1$, $J^2{}_1 = -1$, $J^1{}_2 = -2$, $J^2{}_2 = -0.4$.

$$
\begin{aligned}
\hat{e}_1 \cdot \hat{e}_1 &= J^1{}_1 J^1{}_1 + J^2{}_1 J^2{}_1 = 1 + 1 = 2 \\
\hat{e}_2 \cdot \hat{e}_2 &= J^1{}_2 J^1{}_2 + J^2{}_2 J^2{}_2 = 4 + 0.16 = 4.16 \\
\hat{e}_1 \cdot \hat{e}_2 &= J^1{}_1 J^1{}_2 + J^2{}_1 J^2{}_2 = -2 + 0.4 = -1.6 \\
\hat{e}_2 \cdot \hat{e}_1 &= J^1{}_2 J^1{}_1 + J^2{}_2 J^2{}_1 = -2 + 0.4 = -1.6
\end{aligned}
$$

for (d) $J^1{}_1 = 2$, $J^2{}_1 = -0.7$, $J^1{}_2 = -1$, $J^2{}_2 = 0.8$.

$$
\begin{aligned}
\hat{e}_1 \cdot \hat{e}_1 &= J^1{}_1 J^1{}_1 + J^2{}_1 J^2{}_1 = 4 + 0.49 = 4.49 \\
\hat{e}_2 \cdot \hat{e}_2 &= J^1{}_2 J^1{}_2 + J^2{}_2 J^2{}_2 = 1 + 0.64 = 1.64 \\
\hat{e}_1 \cdot \hat{e}_2 &= J^1{}_1 J^1{}_2 + J^2{}_1 J^2{}_2 = -2 - 0.56 = -2.56 \\
\hat{e}_2 \cdot \hat{e}_1 &= J^1{}_2 J^1{}_1 + J^2{}_2 J^2{}_1 = -2 - 0.56 = -2.56
\end{aligned}
$$

2.3 We Want Objects That Transform According to Coordinate Transformation. Why?

In the field of geometry, it is important to make objects that transform according to the coordinate transformation J and J^{-1}. Why?

2.3.1 Coordinate transformation in n dimension

Consider a coordinate system e_i, $i = 1, \ldots n$, with metric tensor g_{ij} and transformation

$$
\begin{aligned}
\hat{e}_i &= J^k{}_i e_k \\
e_i &= {J^{-1}}^k{}_i \hat{e}_k
\end{aligned}
$$

$$\hat{g} = J^T g J$$

The vector components

$$x = [x^1, \ldots, x^n]$$

are transformed as

$$\hat{x} = J^{-1} x$$

Invariance in vector length can be proven as follows:

$$\begin{aligned}
\|x\|^2 &= (x^i e_i)\cdot(x^j e_j) \\
&= x^i x^j e_i \cdot e_j \\
&= x^i x^j g_{ij} \\
&= J^i{}_k \hat{x}^k J^j{}_s \hat{x}^s ((J^{-1})^T \hat{g} J^{-1})_{ij} \\
&= J^i{}_k J^j{}_s {J^{-1}}^r{}_i {J^{-1}}^t{}_j \hat{g}_{rt} \hat{x}^k \hat{x}^s \\
&= {J^{-1}}^r{}_i J^i{}_k {J^{-1}}^t{}_j J^j{}_s \hat{g}_{rt} \hat{x}^k \hat{x}^s \\
&= \delta^r{}_k \delta^t{}_s \hat{g}_{rt} \hat{x}^k \hat{x}^s \\
&= \hat{g}_{ks} \hat{x}^k \hat{x}^s \\
&= \|\hat{x}\|^2
\end{aligned}$$

2.4 Summary

Definition 2.4.1 Given n types of elements $e_i, i = 1, \ldots, n$, define the binary operations $\cdot$, $+$ and scalar multiplication such that

$$\alpha e_i + \beta e_i = (\alpha + \beta) e_i, \qquad \alpha, \beta \in \mathbb{R}$$

$$e_i \cdot e_j = g_{ij} \qquad g_{ij} \in \mathbb{R}$$

g_{ij} is called the metric tensor.

The distributive rule is applicable, that is,

$$(\alpha e_i + \beta e_j) \cdot \lambda e_k = \alpha \lambda e_i \cdot e_k + \beta \lambda e_j \cdot e_k$$

Lemma 2.4.1 *Given a coordinate transformation $\hat{e}_i = J^k{}_i e_k$, the components of e_k transform as $\hat{x}^i = {J^{-1}}^i{}_k x^k$ with $J^k{}_i {J^{-1}}^i{}_j = \delta^k{}_j$.*

The vector basis transforms in a covariant way, whereas the vector component transforms in a contravariant way. For contravariant components, indices are on top, while for covariant components, the indices are at the bottom.

Lemma 2.4.2 *The metric tensor transform as*

$$\hat{g} = J^T g J$$

Lemma 2.4.3 *The vector length is invariant with respect to the coordinate transformation:*

$$\hat{x} \cdot \hat{x} = x \cdot x$$

Chapter 3

Fields, Derivatives and Curvilinear Coordinates

This chapter discusses functions on $\mathbb{R}^n$ and their differentials. Nonlinear coordinate transformation is also introduced. The concept of basis vectors changing at every point in space is explained thoroughly, and the Christoffel symbols in $\mathbb{R}^n$ are introduced. We also lay the foundations for calculating the length of a path in space. Some reference materials related to the subject discussed in this chapter can be found in Philip R. Wallace [19], Frederick Max Stein [15] and William R. Parzynski and Philip W. Zipse [20].

3.1 Fields and Derivatives

3.1.1 Scalar fields in one-dimensional space

Given $x \in \mathbb{R}$, we define a function that maps every point to a real number as

$$f : \mathbb{R} \mapsto \mathbb{R}$$

Examples

1. $x \in [0, 1], f(x) = \sin(x)$

2. $x \in [-1, 1], f(x) = 3x + 4$

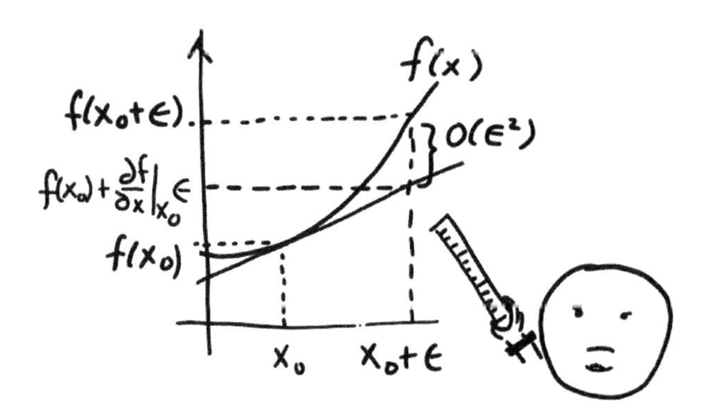

Figure 3.1: The graphical meaning of the first-order Taylor series expansion for $f : \mathbb{R} \mapsto \mathbb{R}$.

3.1.2 Scalar fields in two-dimensional space

Given $(x^1, x^2) \in \mathbb{R} \times \mathbb{R}$, we define a function that maps every point to a real number as

$$f : \mathbb{R} \times \mathbb{R} \mapsto \mathbb{R}$$

Examples

1. $(x^1, x^2) \in [-1, 1]^2, f(x^1, x^2) = \sin(x^1 + 2x^2)$
2. $(x^1, x^2) \in [1, 10]^2, f(x^1, x^2) = x^1 \log(x^2)$
3. The temperature of our table top.
4. Consider that green beans cost \$1 per 100 grams and red beans cost \$1.10 per 100 grams. In the space of all bags containing green and red beans, the total cost of a mixture of green and red beans is a scalar field.

3.1.3 Derivatives

Given a scalar field (function) in one-dimensional space, $f : \mathbb{R} \mapsto \mathbb{R}$, its derivative is defined as

$$\frac{df(x)}{dx} = \lim_{\epsilon \to 0} \frac{f(x + \epsilon) - f(x)}{\epsilon}$$

Note that (df/dx) varies at different points x. In two-dimensional space, we consider a basis, $x = x^1 e_1 + x^2 e_2 = x^i e_i$ with $g_{ij} = \delta_{ij}$. Then, we can

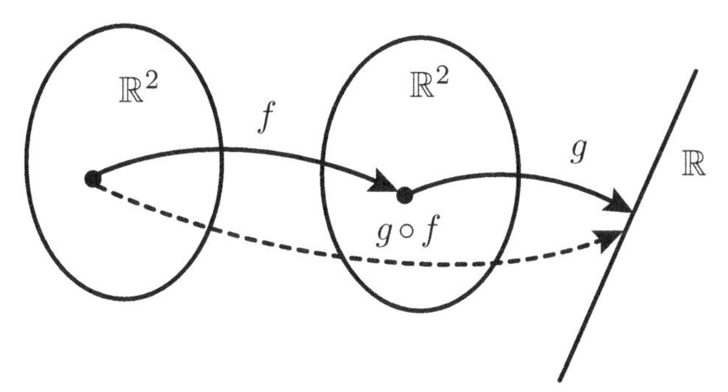

Figure 3.2: Graphical representation of the composite of functions f and g.

define the derivatives separately in these two directions:

$$\begin{aligned}\frac{\partial f(x^1, x^2)}{\partial x^1} &= \lim_{\epsilon \to 0} \frac{f(x^1 + \epsilon, x^2) - f(x^1, x^2)}{\epsilon} \\ \frac{\partial f(x^1, x^2)}{\partial x^2} &= \lim_{\epsilon \to 0} \frac{f(x^1, x^2 + \epsilon) - f(x^1, x^2)}{\epsilon}\end{aligned}$$

Note that the values of the differentials vary at different points: $x = (x^1, x^2)$. The above equation can be generalized to define a derivative in some direction $v = v^1 e_1 + v^2 e_2$. This is called a direction derivative:

$$\frac{\partial f(x)}{\partial v} = \lim_{\epsilon \to 0} \frac{f(x + \epsilon v) - f(x)}{\epsilon}$$

Composite function

Consider two functions,

$$f : \mathbb{R}^2 \mapsto \mathbb{R}^2 \text{ and } g : \mathbb{R}^2 \mapsto \mathbb{R}$$

We can pass $x \in \mathbb{R}^2$ into f to get $f(x)$. We can chain functions together by passing the output of f as input to g. Then, $g(f(x)) \in \mathbb{R}$. This is called a composition function, which we denote as $g \circ f$. Figure 3.2 illustrates a graphical representation of a composite function.

Product rule, chain rule and Taylor series

Let $x \in \mathbb{R}$, $f : \mathbb{R} \mapsto \mathbb{R}$ and $g : \mathbb{R} \mapsto \mathbb{R}$. Then, the product rule gives

$$\frac{d}{dx} f(x)g(x) = \frac{df(x)}{dx} g(x) + f(x) \frac{dg(x)}{dx}$$

For the chain rule, let $y = f(x)$. Then,

$$\frac{dg(f(x))}{dx} = \frac{dg(y)}{dx} = \frac{dg(y)}{dy}\frac{dy}{dx} = \frac{dg(y)}{dy}\frac{df(x)}{dx}$$

The Taylor series gives

$$f(x_0 + \epsilon) = f(x_0) + \left.\frac{\partial f}{\partial x}\right|_{x_0} \epsilon + \left.\frac{\partial^2 f}{\partial x^2}\right|_{x_0} \frac{\epsilon^2}{2} + \cdots$$

The Taylor series for multivariate functions is $x = x^i e_i$ and $\epsilon = \epsilon^i e_i$,

$$\begin{aligned} f(x+\epsilon) &= f(x) + \sum_i \frac{\partial f}{\partial x^i}\epsilon^i + O(\epsilon^2) \\ &= f(x) + \frac{\partial f}{\partial x^i}\epsilon^i + O(\epsilon^2) \end{aligned} \tag{3.1}$$

Einstein's notation for summation over double indices is used here. Figure 3.1 illustrates how the above equation (EQ. (3.1)) can be interpreted in a geometrical way.

Exercise 3.1.1

1. Show that

$$\frac{\partial f(x)}{\partial v} = \sum_i v^i \frac{\partial f}{\partial x^i}$$

Solution:

$$\begin{aligned} \lim_{\epsilon \to 0} \frac{f(x+\epsilon v) - f(x)}{\epsilon} &= \lim_{\epsilon \to 0} \frac{f(x^1 + \epsilon v^1, x^2 + \epsilon v^2) - f(x^1, x^2)}{\epsilon} \\ &= \lim_{\epsilon \to 0} \frac{\epsilon v^1(\partial_{x^1} f) + \epsilon v^2(\partial_{x^2} f) + O(\epsilon^2)}{\epsilon} \\ &= v^1 \frac{\partial f}{\partial x^1} + v^2 \frac{\partial f}{\partial x^2} \end{aligned} \tag{3.2}$$

---

2. How does $(\partial/\partial x^1, \partial/\partial x^2)$ transform into $(\partial/\partial \hat{x}^1, \partial/\partial \hat{x}^2)$? Let e_1, e_2 and $\hat{e}_1, \hat{e}_2$ be related by

$$\hat{e}_i = J^k{}_i e_k$$

Solution: Consider a point $x = x^i e_i = \hat{x}^i \hat{e}_i$. The same point in space is represented by two coordinate systems. Then, the value of the function on x is the same regardless of which coordinate system is used:

$$f(x^i e_i) = f(\hat{x}^i \hat{e}_i)$$

$$\begin{aligned}
\partial_{\hat{x}^i} f &= \lim_{\epsilon \to 0} \frac{f(\hat{x}^j \hat{e}_j + \epsilon \hat{e}_i) - f(\hat{x}^j \hat{e}_j)}{\epsilon} \\
&= \lim_{\epsilon \to 0} \frac{f(x^j e_j + \epsilon J^j{}_i e_j) - f(x^j e_j)}{\epsilon}
\end{aligned}$$

Use multivariate Taylor series expansion up to first order:

$$\begin{aligned}
\partial_{\hat{x}^i} f &= \lim_{\epsilon \to 0} \frac{f((x^j + \epsilon J^j{}_i) e_j) - f(x^j e_j)}{\epsilon} \\
&= \lim_{\epsilon \to 0} \frac{1}{\epsilon} \left(f(x^j e_j) + \frac{\partial f}{\partial x^j} \epsilon^j J^i{}_j - f(x^j e_j) + O(\epsilon^2) \right) \\
&= \lim_{\epsilon \to 0} \frac{\epsilon J^j{}_i \partial_{x^j} f + O(\epsilon^2)}{\epsilon} = J^j{}_i \partial_{x^j} f
\end{aligned}$$

Therefore,

$$\partial_{\hat{x}^i} = J^j{}_i \partial_{x^j}$$

The key takeaway here is that partial derivatives transform the same way as the basis vectors. We say that partial derivatives transform in a covariant way. Recall that the components of basis vectors transform in a contravariant way.

3.1.4 Vector fields

Similar to scalar fields, vector fields map every point in the underlying space to a vector. Given a basis e_i, $i = 1, \ldots, n$, use this same basis as basis vectors for the vector fields.

Example

1. For a two-dimensional space with a basis, e_1, e_2, and a function, $f : \mathbb{R} \times \mathbb{R} \mapsto \mathbb{R}$, a vector field can be generated using the differential $(\partial f / \partial x^1, \ldots, \partial f / \partial x^n)$.

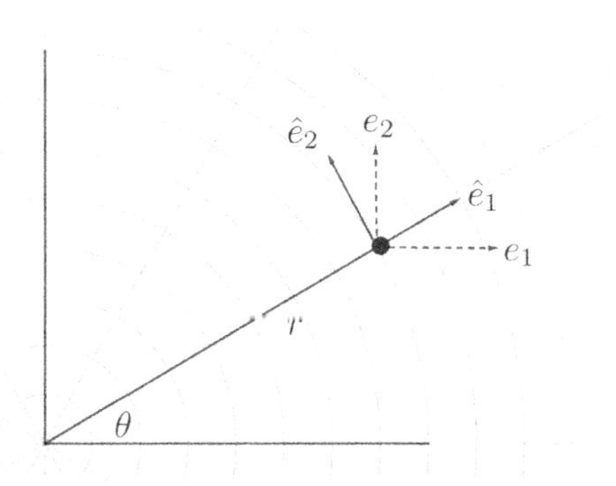

Figure 3.3: Cartesian to polar coordinate transformation.

3.2 Nonlinear Coordinate Transformation

In the previous chapters, we studied coordinate transformations that are the same throughout the underlying space. We illustrated this with examples in two-dimensional space. The governing equation for coordinate transformation is given by

$$\hat{e}_i = J^j{}_i e_j \tag{3.3}$$

where $J^j{}_i$ is a constant over the whole space. Let $x = x^i e_i = \hat{x}^i \hat{e}_i$ be a point in the underlying space. Consider a transformation over this space where $J^j{}_i = J^j{}_i(x)$ depends on the point in space.

$$\hat{e}_i = J^j{}_i(x) e_j$$

How can we then write out the correct transformation?

3.2.1 Cartesian to polar coordinates

Let's run through an example to illustrate this concept. Consider the two-dimensional Cartesian $(x^1, x^2) = (q^1, q^2)$ and polar coordinates $(\hat{x}^1, \hat{x}^2) = (r, \theta)$ as illustrated in Figure 3.3.

$$\begin{aligned} r^2 &= (q^1)^2 + (q^2)^2 \\ \theta &= \tan^{-1}\left(q^2/q^1\right) \\ q^1 &= r\cos(\theta) \\ q^2 &= r\sin(\theta) \end{aligned} \tag{3.4}$$

A transformation of variables is not enough to determine basis vectors. We need another condition, which is to specify that the new basis vectors are vectors that point in the direction of small changes in the new variables. Given a point $x = q^1 e_1 + q^2 e_2$,

$$\begin{aligned} \hat{e}_1 &= \frac{\partial x}{\partial r} \\ \hat{e}_2 &= \frac{\partial x}{\partial \theta} \end{aligned} \tag{3.5}$$

$\hat{e}_1, \hat{e}_2$ are evaluated to be

$$\begin{aligned} \hat{e}_1 &= \cos\theta e_1 + \sin\theta e_2 \\ \hat{e}_2 &= -r\sin\theta e_1 + r\cos\theta e_2 \end{aligned} \tag{3.6}$$

Then,

$$J(\theta) = \begin{pmatrix} \frac{\partial q^1}{\partial r} & \frac{\partial q^1}{\partial \theta} \\ \frac{\partial q^2}{\partial r} & \frac{\partial q^2}{\partial \theta} \end{pmatrix} = \begin{pmatrix} \cos\theta & -r\sin\theta \\ \sin\theta & r\cos\theta \end{pmatrix}$$

To verify,

$$\begin{pmatrix} \cos\theta & \sin\theta \\ -r\sin\theta & r\cos\theta \end{pmatrix} \begin{pmatrix} e_1 \\ e_2 \end{pmatrix} = \begin{pmatrix} \cos\theta e_1 + \sin\theta e_2 \\ -r\sin\theta e_1 + r\cos\theta e_2 \end{pmatrix} \tag{3.7}$$

In general,

$$x^i = x^i(\hat{x}^1, \ldots, \hat{x}^n)$$

The Jacobian is given by

$$J^j{}_i = \frac{\partial x^j}{\partial \hat{x}^i}$$

Then, the transformation rule becomes

$$\hat{e}_i = J^j{}_i e_j \tag{3.8}$$

Can we calculate the length of a position vector in polar coordinates? In Cartesian coordinates, we have that the square of length is given by $r^2 = (q^1)^2 + (q^2)^2$. The same point in space in polar coordinates is (r, θ). If we use the metric tensor equation for calculating length,

$$\begin{pmatrix} r & \theta \end{pmatrix} \cdot \begin{pmatrix} 1 & 0 \\ 0 & r^2 \end{pmatrix} \cdot \begin{pmatrix} r \\ \theta \end{pmatrix} = r^2(1 + \theta^2)$$

The above result is obviously wrong. We know the correct result is r^2. What happened?

Exercise 3.2.1

1. Show that Eq. (3.5) leads to Eq. (3.6).

   Solution: The equations can be derived by differentiating Eq. (3.4) directly.

2. Draw diagrams to verify that Eq. (3.6) is correct.

   Solution: Figure 3.3 illustrates that the angles between $\hat{e}_1$ and e_1 and between $\hat{e}_2$ and e_2 are θ. Use sine and cosine to resolve the components between the vectors.

3. Show that Eq. (3.8) is correct for the Cartesian to polar coordinates transformation

   Solution: By substituting expressions for polar coordinates into the Jacobian, we recover Eq. (3.6).

4. What is the metric tensor in polar coordinates for Euclidean space?

   Solution: First, obtain the basis vectors $\hat{e}_1$ and $\hat{e}_2$ and then perform the dot product using the metric tensor of the Cartesian coordinate's basis vectors $e_i \cdot e_j = \delta_{ij}$:

$$\begin{aligned}
\hat{g}_{11} = \hat{e}_1 \cdot \hat{e}_1 &= (\cos\theta e_1 + \sin\theta e_2) \cdot (\cos\theta e_1 + \sin\theta e_2) = 1 \\
\hat{g}_{12} = \hat{e}_1 \cdot \hat{e}_2 &= (\cos\theta e_1 + \sin\theta e_2) \cdot (-r\sin\theta e_1 + r\cos\theta e_2) = 0 \\
\hat{g}_{22} = \hat{e}_2 \cdot \hat{e}_2 &= (-r\sin\theta e_1 + r\cos\theta e_2) \cdot (-r\sin\theta e_1 + r\cos\theta e_2) = r^2
\end{aligned}$$

   Therefore,

$$\hat{g} = \begin{pmatrix} 1 & 0 \\ 0 & r^2 \end{pmatrix}$$

5. Consider that the spherical coordinate representation of space-time is defined on $\mathbb{R}^4$ as

$$\begin{aligned}
q^0 &= t \\
q^1 &= r\sin\theta\cos\phi \\
q^2 &= r\sin\theta\sin\phi \\
q^3 &= r\cos\theta
\end{aligned} \tag{3.9}$$

The metric tensor in Cartesian coordinates for the space-time geometry is given by the Minkowski metric

$$g = \begin{pmatrix} 1 & 0 & 0 & 0 \\ 0 & -1 & 0 & 0 \\ 0 & 0 & -1 & 0 \\ 0 & 0 & 0 & -1 \end{pmatrix} \tag{3.10}$$

What is the metric tensor for polar coordinates?

Solution: First, we compute the basis vectors in this new coordinate representation:

$$\begin{aligned} q^0 &= t \\ q^1 &= r\sin\theta\cos\phi \\ q^2 &= r\sin\theta\sin\phi \\ q^3 &= r\cos\theta \end{aligned} \tag{3.11}$$

$x = q^0e_0 + q^1e_1 + q^2e_2 + q^3e_3$,

$$\begin{aligned} \hat{e}_0 &= \frac{\partial x}{\partial t} = e_0 \\ \hat{e}_1 &= \frac{\partial x}{\partial r} = \sin\theta\cos\phi e_1 + \sin\theta\sin\phi e_2 + \cos\theta e_3 \\ \hat{e}_2 &= \frac{\partial x}{\partial \theta} = r\cos\theta\cos\phi e_1 + r\cos\theta\sin\phi e_2 + -r\sin\theta e_3 \\ \hat{e}_3 &= \frac{\partial x}{\partial \phi} = -r\sin\theta\sin\phi e_1 + r\sin\theta\cos\phi e_2 \end{aligned} \tag{3.12}$$

Porforming inner products,

$$\begin{aligned} \hat{e}_0 \cdot \hat{e}_i &= \delta_{0i} \\ \hat{e}_1 \cdot \hat{e}_1 &= -1 \\ \hat{e}_1 \cdot \hat{e}_2 &= 0 \\ \hat{e}_1 \cdot \hat{e}_3 &= 0 \\ \hat{e}_2 \cdot \hat{e}_2 &= -r^2 \\ \hat{e}_2 \cdot \hat{e}_3 &= 0 \\ \hat{e}_3 \cdot \hat{e}_3 &= -r^2\sin^2\theta \end{aligned} \tag{3.13}$$

The metric tensor is hence

$$g = \begin{pmatrix} 1 & 0 & 0 & 0 \\ 0 & -1 & 0 & 0 \\ 0 & 0 & -r^2 & 0 \\ 0 & 0 & 0 & -r^2\sin^2\theta \end{pmatrix} \tag{3.14}$$

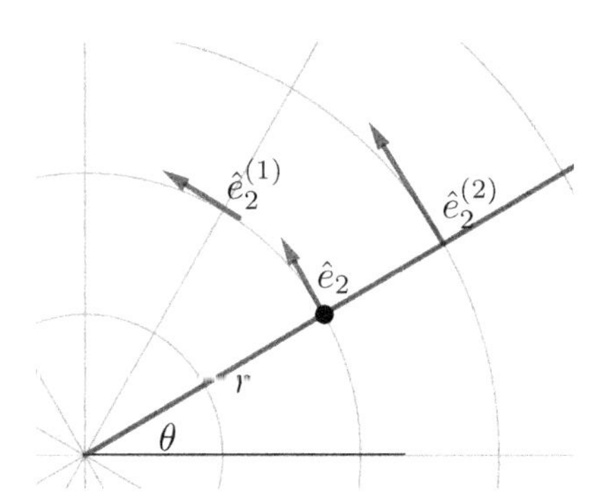

Figure 3.4: $\hat{e}_2$ changes at different positions in polar coordinates. $\hat{e}_2^{(1)}$ is longer than $\hat{e}_2$ and $\hat{e}_2^{(2)}$ is pointing in a different direction from $\hat{e}_2$.

3.3 Christoffel Symbols in Polar Coordinates

Consider polar coordinates, and the magnitude of the basis vector $\hat{e}_2$ increases with r, i.e. $|\hat{e}_2| = r$. The direction of $\hat{e}_2$ also changes with θ, as illustrated in Figure 3.4. At a point x,

$$\begin{aligned} \frac{\partial \hat{e}_2}{\partial r} &= \hat{e}_2/r \\ \frac{\partial \hat{e}_2}{\partial \theta} &= -r\hat{e}_1 \end{aligned} \tag{3.15}$$

The above equation states that the rate of change of $\hat{e}_2$ when r changes slightly is $\hat{e}_2/r$. One can express this rate of change in terms of the original basis (before the change). We consider the rate of change at the same point x; this rate of change is also a vector at the point x. Since it is a vector at the point x, one can expand this vector in the original basis $\hat{e}_2, \hat{e}_1$ at the point x:

$$\frac{\partial \hat{e}_2}{\partial r} = \alpha\hat{e}_1 + \beta\hat{e}_2$$

For some α and β given by

$$\begin{aligned} \frac{\partial \hat{e}_2}{\partial r} \cdot \hat{e}_1 &= \alpha\hat{e}_1 \cdot \hat{e}_1 + \beta\hat{e}_2 \cdot \hat{e}_1 \\ &= \alpha \\ \frac{\partial \hat{e}_2}{\partial r} \cdot \hat{e}_2 &= \alpha\hat{e}_1 \cdot \hat{e}_2 + \beta\hat{e}_2 \cdot \hat{e}_2 \\ &= \beta r^2 \end{aligned}$$

$$\begin{aligned} \alpha &= \frac{\partial \hat{e}_2}{\partial r} \cdot \hat{e}_1 \\ \beta &= \frac{1}{r^2} \frac{\partial \hat{e}_2}{\partial r} \cdot \hat{e}_2 \end{aligned}$$

From Eq. (3.15),

$$\begin{aligned} \alpha &= 0 \\ \beta &= 1/r \end{aligned}$$

Since there are many combinations, we will soon run out of mathematical symbols for each symbol to represent one combination. So, we use one symbol with indices:

$$\begin{aligned} \frac{\partial \hat{e}_1}{\partial r} &= \Gamma^1{}_{11}\hat{e}_1 + \Gamma^2{}_{11}\hat{e}_2 \\ \frac{\partial \hat{e}_1}{\partial \theta} &= \Gamma^1{}_{12}\hat{e}_1 + \Gamma^2{}_{12}\hat{e}_2 \\ \frac{\partial \hat{e}_2}{\partial r} &= \Gamma^1{}_{21}\hat{e}_1 + \Gamma^2{}_{21}\hat{e}_2 \\ \frac{\partial \hat{e}_2}{\partial \theta} &= \Gamma^1{}_{22}\hat{e}_1 + \Gamma^2{}_{22}\hat{e}_2 \end{aligned} \tag{3.16}$$

Now, we are ready to write equations in a general form:

$$\frac{\partial e_i}{\partial x^j} = \Gamma^k{}_{ij} e_k \tag{3.17}$$

$\Gamma^k{}_{ij}$ are called the Christoffel symbols. Note that this equation is derived on a flat space, and we have not shown that it is valid for a curved space. Since we have not yet discussed curved spaces, we will differ from further discussing the Christoffel symbols in curved spaces in a later part of this book.

Exercise 3.3.1

1. Make the Christoffel symbol the subject in Eq. (3.17).

   Solution: Perform the dot product on both sides of the

above-mentioned equation using another basis vector:

$$\begin{aligned}
\Gamma^k{}_{ij} e_k &= \frac{\partial e_i}{\partial x^j} \\
\Gamma^k{}_{ij} e_k \cdot e_l &= \frac{\partial e_i}{\partial x^j} \cdot e_l \\
\Gamma^k{}_{ij} g_{kl} &= \frac{\partial e_i}{\partial x^j} \cdot e_l \\
\Gamma^k{}_{ij} g_{kl} g^{lm} &= \frac{\partial e_i}{\partial x^j} \cdot e_l g^{lm} \\
\Gamma^k{}_{ij} \delta_k{}^m &= \frac{\partial e_i}{\partial x^j} \cdot e_l g^{lm} \\
\Gamma^m{}_{ij} &= \frac{\partial e_i}{\partial x^j} \cdot e_l g^{lm}
\end{aligned} \tag{3.18}$$

---

2. Work out all the Christoffel symbols in two-dimensional polar coordinates.

   Solution: Using Eq. (3.16), evaluate the differentials for the basis vectors:

$$\begin{aligned}
\frac{\partial \hat{e}_1}{\partial r} &= 0 \\
\frac{\partial \hat{e}_1}{\partial \theta} &= \frac{\hat{e}_2}{r} \\
\frac{\partial \hat{e}_2}{\partial \theta} &= -r\hat{e}_1
\end{aligned}$$

   Substituting into Eq. (3.16),

$$\begin{aligned}
\Gamma^1{}_{11}\hat{e}_1 + \Gamma^2{}_{11}\hat{e}_2 &= 0 \\
\Gamma^1{}_{12}\hat{e}_1 + \Gamma^2{}_{12}\hat{e}_2 &= \frac{\hat{e}_2}{r} \\
\Gamma^1{}_{22}\hat{e}_1 + \Gamma^2{}_{22}\hat{e}_2 &= -r\hat{e}_1
\end{aligned}$$

   Since $\hat{e}_1 \cdot \hat{e}_2 = 0$,

$$\begin{aligned}
\Gamma^1{}_{11} = \Gamma^2{}_{11} = \Gamma^1{}_{12} = \Gamma^1{}_{21} = \Gamma^2{}_{22} &= 0 \\
\Gamma^2{}_{12} = \Gamma^2{}_{21} = \frac{1}{r}\Gamma^1{}_{22} &= -r
\end{aligned}$$

---

3. Work out all the Christoffel symbols in 2D Cartesian coordinates.
   Solution: Christoffel symbols are all zero for Cartesian coordinates in any dimension.

---

4. Prove that the Christoffel symbols are symmetric with respect to the lower indices for polar coordinates:

$$\Gamma^k{}_{ij} = \Gamma^k{}_{ji}$$

Solution: For polar coordinate transformations in Cartesian coordinates, we need to prove

$$\frac{\partial \hat{e}_i}{\partial \hat{x}^j} = \frac{\partial \hat{e}_j}{\partial \hat{x}^i}$$

Since,

$$\hat{e}_i = \frac{\partial x^k}{\partial \hat{x}^i} e_k$$

$$\frac{\partial \hat{e}_i}{\partial \hat{x}^j} = \frac{\partial^2 x^k}{\partial \hat{x}^j \partial \hat{x}^i} e_k = \frac{\partial \hat{e}_j}{\partial \hat{x}^i}$$

3.4 Calculating Path Lengths

First, we learn the concept of path length parameterization. We want to trace a path in 2D space using a variable t, with $t = 0$ at the starting point and $t = 1$ when the path ends. In Cartesian coordinates, we can write

$$\begin{aligned} q^1 &= q^1(t) \\ q^2 &= q^2(t) \end{aligned}$$

As illustrated in Figure 3.5, the total length of a path is the sum or integral of small line segments. The path length is then given by

$$\int_0^1 \sqrt{(\dot{q}^1(t), \dot{q}^2(t)) \cdot (\dot{q}^1(t), \dot{q}^2(t))}dt \tag{3.19}$$

This equation simply states that the total path length is the sum of all infinitesimal lengths of differentials in the required coordinate system. Note that the dot product $()\cdot()$ involves the metric tensor.

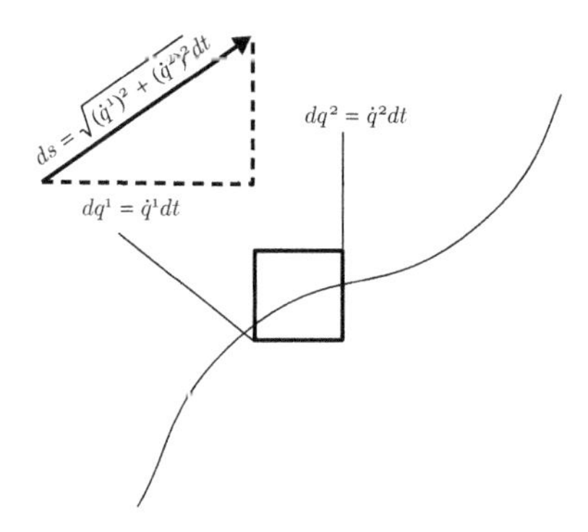

Figure 3.5: The length of a path is the sum of small segments $ds^2 = (dq^1)^2 + (dq^2)^2$, $\int ds = \int \sqrt{(\dot{q}^1)^2 + (\dot{q}^2)^2}dt$.

Exercise 3.4.1

1. Draw a path parameterized by the following equations:

$$\begin{aligned} q^1(t) &= 2t \\ q^2(t) &= -t \\ t &\in [0,1] \end{aligned} \tag{3.20}$$

Solution: This is a path of the straight line $q^1 = -2q^2$.

---

2. Draw a path parameterized by the following equations:

$$\begin{aligned} q^1(t) &= t^2 + 0.1 \\ q^2(t) &= t \\ t &\in [0,1] \end{aligned} \tag{3.21}$$

Solution: Use a parametric plotting software to plot out the path. The equation of the path is $q^1 = (q^2)^2 + 0.1$.

---

3. Calculate the path length given by Eq. (3.20) using Cartesian coordinates, and repeat the exercise using polar coordinates.

Solution: For Eq. (3.20),

$$\begin{aligned} v^1 = \frac{dq^1(t)}{dt} &= 2 \\ v^2 = \frac{dq^2(t)}{dt} &= -1 \\ g_{ij} &= \delta_{ij} \\ (v^1, v^2) \cdot (v^1, v^2) &= v^i g_{ij} v^j = 5 \end{aligned}$$

The curve length is

$$l = \int_0^1 \sqrt{5} dt = \sqrt{5}$$

In polar coordinates,

$$\begin{aligned} r(t) &= \sqrt{5}t \\ \theta(t) &= \arctan(-1/2) \approx -0.464 \\ v^1 &= \frac{dr(t)}{dt} = \sqrt{5} \\ v^2 &= \frac{d\theta(t)}{dt} = 0 \\ (v^1, v^2) \cdot (v^1, v^2) &= v^i g_{ij} v^j = 5 \end{aligned}$$

Therefore, the curve length is $\sqrt{5}$.

$$l = \int_0^1 \sqrt{5} dt = \sqrt{5}$$

If we don't use integration and instead compute the path length using the metric tensor dot product alone,

$$l = \begin{pmatrix} \sqrt{5} & 0.464 \end{pmatrix} \begin{pmatrix} 1 & 0 \\ 0 & 5 \end{pmatrix} \begin{pmatrix} \sqrt{5} \\ 0.464 \end{pmatrix} = 5(1 + 0.464)$$

which is the wrong answer.

---

4. Calculate the path length given by Eq. (3.21) using Cartesian coordinates, and repeat the exercise using polar coordinates.

   Solution: For the path length of Eq. (3.21) in Cartesian coordinates,

$$\begin{aligned} q^1(t) &= t^2 + 0.1 \\ q^2(t) &= t \\ v^1 &= \frac{dq^1(t)}{dt} = 2t \\ v^2 &= \frac{dq^2(t)}{dt} = 1 \\ (v^1, v^2) \cdot (v^1, v^2) &= v^i g_{ij} v^j = 4t^2 + 1 \end{aligned}$$

The curve length is

$$l = \int_0^1 \sqrt{4t^2 + 1} dt = 1.478\ldots$$

In polar coordinates,

$$\begin{aligned} r(t) &= \sqrt{(t^2 + 0.1)^2 + t^2} \\ \theta(t) &= \arctan\left(\frac{t}{t^2 + 0.1}\right) \\ v^1 &= \frac{dr(t)}{dt} = \frac{2(t^3 + 0.6t)}{r(t)} \\ v^2 &= \frac{d\theta(t)}{dt} = \frac{0.1 - t^2}{r(t)^2} \\ (v^1, v^2) \cdot (v^1, v^2) &= v^i g_{ij} v^j = v^1 v^1 + r^2 v^2 v^2 \\ &= \frac{4(t^3 + 0.6t)^2 + (0.1 - t^2)^2}{r^2} \end{aligned}$$

The curve length is

$$l = \int_0^1 \sqrt{v^1 v^1 + r^2 v^2 v^2} dt = 1.478...$$

Attempting to simplify,

$$\frac{4(t^3 + 0.6t)^2 + (0.1 - t^2)^2}{r^2}$$

This equation can be factorized to get $4t^2 + 1$:

$$\frac{0.01(4t^2 + 1)(100t^4 + 120t^2 + 1)}{r^2} = 4t^2 + 1$$

After factorizing, one factor in the denominator and numerator cancels to give $4t^2 + 1$.

3.4.1 How does the position vector trace through polar coordinates?

In the rectilinear coordinate system, we trace from the origin using the coefficients and move along the respective basis vector. Eventually, we reach the point. In polar coordinates, how does this happen?

3.5 Summary

Definition 3.5.1 Given a differentiable mapping $\hat{x} = \hat{x}(x)$, the Jacobian is defined as

$$J^j{}_i = \frac{\partial x^j}{\partial \hat{x}^i}$$

Definition 3.5.2 Polar coordinates defined on $\mathbb{R}^2$ are given by

$$\begin{aligned} q^1 &= r\cos\theta \\ q^2 &= r\sin\theta \end{aligned}$$

Using the Jacobian, the transformations between basis vectors are

$$\begin{aligned} \hat{e}_1 &= \cos\theta e_1 + \sin\theta e_2 \\ \hat{e}_2 &= -r\sin\theta e_1 + r\cos\theta e_2 \end{aligned} \tag{3.22}$$

The metric tensor in polar coordinate representation is given by

$$\hat{g} = \begin{pmatrix} \hat{e}_1 \cdot \hat{e}_1 & \hat{e}_1 \cdot \hat{e}_2 \\ \hat{e}_2 \cdot \hat{e}_1 & \hat{e}_2 \cdot \hat{e}_2 \end{pmatrix} = \begin{pmatrix} 1 & 0 \\ 0 & r^2 \end{pmatrix}$$

Definition 3.5.3 The arc length of a path parameterized by $t \in \mathbb{R}$ is given by

$$l = \int_0^1 \sqrt{\frac{dx}{dt} \cdot \frac{dx}{dt}}\, dt$$

where the inner product involves using the metric tensor.

Definition 3.5.4 The Christoffel symbol is defined as

$$\frac{\partial e_i}{\partial x^j} = \Gamma^k{}_{ij} e_k$$

Lemma 3.5.1 The Christoffel symbols is symmetric with respect to the lower indices:

$$\Gamma^k{}_{ij} = \Gamma^k{}_{ji}$$

Chapter 4

Introduction to Curved Spaces Embedded in $\mathbb{R}^n$

In the previous chapters, we learned that the metric tensor is essential to calculating the lengths of vectors. The Jacobian is the central tool for writing one set of basis vectors in terms of another. We learned about the Jacobian transformations between two basis vectors of the same dimension. We use these tools to define a set of points in a curved space embedded in $\mathbb{R}^n$. Let the set of points in curved space (e.g. a surface) be denoted by M. This curved space is also called a manifold.

4.1 Understanding a Surface in $\mathbb{R}^n$

A manifold can be interpreted as a "surface" in high-dimensional space. Sometimes (but not always), we can specify a manifold in $\mathbb{R}^n$ by using level-sets:

$$M = \{x \in \mathbb{R}^n | h_i(x) = 0, i = 1, 2, \dots n - d\}$$

The dimension of M is given by d. h_i are the level-set functions. For example, in $\mathbb{R}^3$, we specify all the points, $(x^1)^2 + (x^2)^2 + (x^3)^2 - 1 = 0$. This defines a sphere. There are certain technicalities with using level-sets to specify manifolds; for example, $h_1(x) = 1$ and $h_2(x) = 0$ will make M an empty set. We shall not go into the details of how to define a manifold; interested readers can refer to one of many differential geometry books for a formal definition of a manifold.

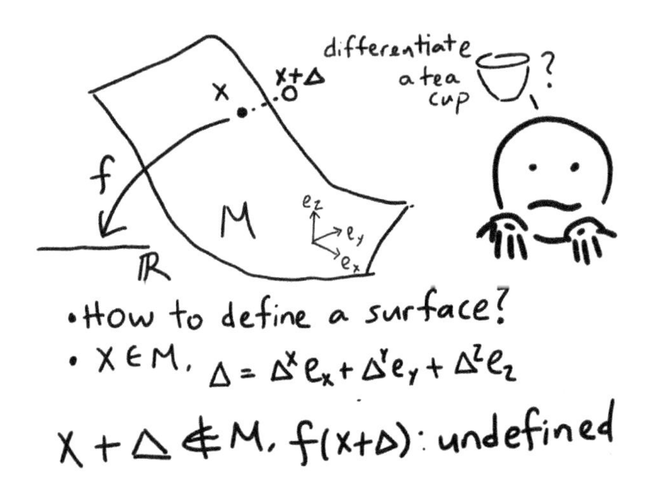

Figure 4.1: Illustration of the issues with defining a function on surfaces when a small displacement brings a point out of the surface.

4.1.1 Scalar and vector fields in M

We define a scalar field as

$$f : M \to \mathbb{R}$$

The scalar field simple states that, for every point in M, we give a scalar value. For example, the temperature on the surface of an airplane. We define vector fields as

$$X : M \to \mathbb{R}^d$$

For every point in M, there is a vector. For example, consider the direction of heat flow on the surface of an airplane. There are some critical issues here:

1. There is a problem with the differentials of the scalar function f on M because it is defined only on M. In the usual calculus in $\mathbb{R}^n$, differentials are computed by taking small steps in all directions in $\mathbb{R}^n$. This does not apply to differentials on functions defined only on M. When we are on M and we take small steps in $\mathbb{R}^n$, we will "step out" of the manifold. The function f is undefined outside of M. Figure 4.1 illustrates that when we take a small step Δ from x to $x + \Delta$, that brings us to a point outside of M.

2. Since we want the vector to lie on M (not pointing out of M), we face problems as soon as we want to compute using these vectors because, for computing, we need a coordinate system. Since the surface is not

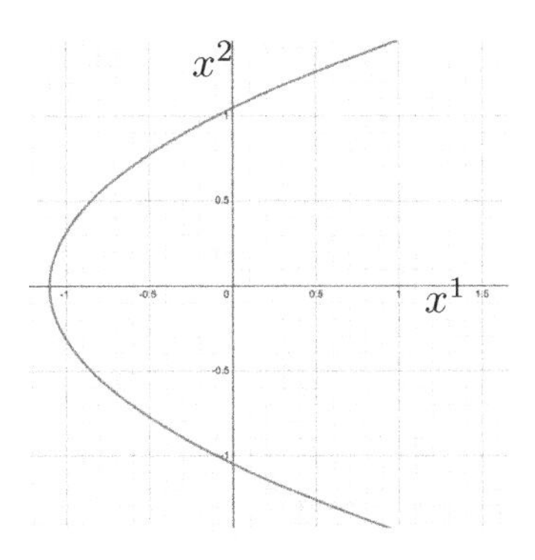

Figure 4.2: Parametric curve for Eq. (4.1).

flat, how do we make a coordinate system that applies to all points in M?

3. We want to perform differentiation and integration on scalars and vectors on M. Without a proper coordinate system, we cannot take limits and therefore cannot perform differentiation and integration.

4.1.2 Concept of parameterization

We came across parameterization in the previous chapter. For example, Figure 4.2 shows the path for the parameterization

$$\begin{aligned} x^1(t) &= t^2 - 1.1 \\ x^2(t) &= t \end{aligned} \tag{4.1}$$

Here, note that we use one parameter to represent two variables. Indeed, this is one way to reduce two dimensions into one dimension.

Exercise 4.1.1

1. What is the shape of the object M parameterized by

$$\begin{aligned} x^1(t) &= 2\cos(2\pi t) \\ x^2(t) &= 3\sin(2\pi t) \end{aligned}$$

Solution: The trace of (x^1, x^2) forms an ellipse. Use a plotting software to verify.

---

2. What is the shape of the object M parameterized by

$$\begin{aligned} x^1(\theta,\phi) &= \cos\phi\sin\theta \\ x^2(\theta,\phi) &= \sin\phi\sin\theta \\ x^3(\theta,\phi) &= \cos\theta \end{aligned}$$

$\phi \in [0, 2\pi)$, $\theta \in (0, \pi)$.

a. Show that $(x^1)^2 + (x^2)^2 + (x^3)^2 = 1$.

b. What path is being traced out for the set of points:

i. $\theta = \pi/2$, $\phi \in [0, 2\pi)$;

ii. $\theta \in (0, \pi)$, $\phi = 0$.

Solution: Use the identity $\cos^2\theta + \sin^2\theta = 1$:

$$\begin{aligned} (x^1)^2 + (x^2)^2 + (x^3)^2 &= \cos^2\phi\sin^2\theta + \sin^2\phi\sin^2\theta + \cos^2\theta \\ &= \sin^2\theta + \cos^2\theta = 1 \end{aligned}$$

Figure 4.3 shows the paths being traced out in this question.

---

3. What is the shape of the object M parameterized by

$$\begin{aligned} x^1(z,\phi) &= \cos\phi \\ x^2(z,\phi) &= \sin\phi \\ x^3(z,\phi) &= z \end{aligned} \tag{4.2}$$

$\phi \in [0, 2\pi)$, $z \in \mathbb{R}$.

a. What path is being traced out for the set of points:

i. $\phi = 0$, $z \in [-1, 1]$;

ii. $\phi \in [0, 2\pi)$, $z = 0$.

Solution: The shape is a cylinder, as shown in Figure 4.4. This figure also shows the trace of the paths $\phi = 0$, $z \in [-1, 1]$ and $\phi \in [0, 2\pi)$, $z = 0$.

4.2 Coordinate System for Points in M

In Exercise 4.1.1, we parameterize the object M using θ, ϕ or t. Indeed, the objects M described by these parameterizations are ellipses, spheres and cylinders. A general manifold embedded in a higher-dimensional space can be parameterized by some functions where the position vectors of the set of points in M are given by a chart.

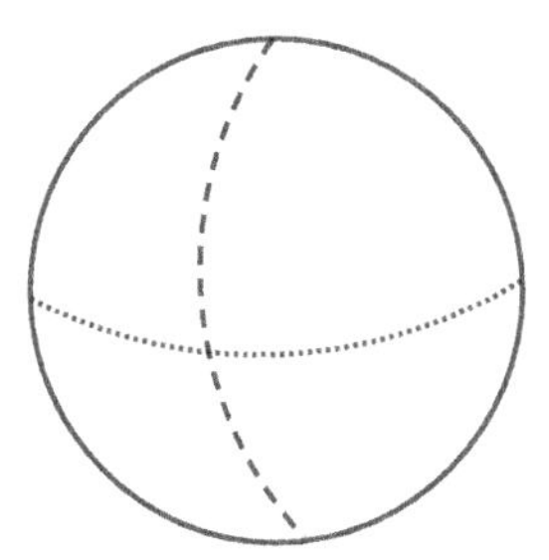

Figure 4.3: The trace of the path along $\theta = \pi/2$, $\phi \in [0, 2\pi)$ (dotted line) and $\theta \in (0, \pi)$, $\phi = 0$ (dashed line) are given as latitudes and longitudes.

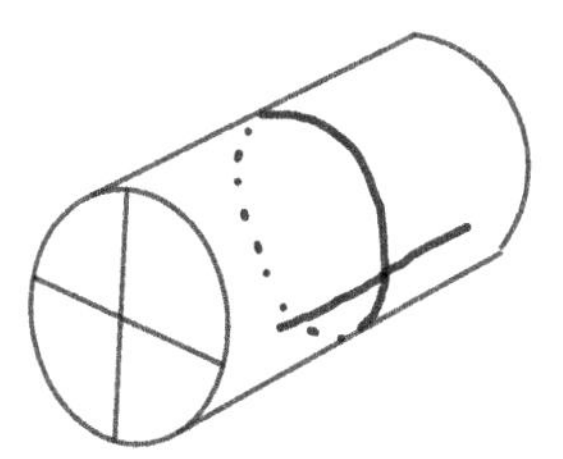

Figure 4.4: The trace of the paths along $\phi = 0$, $z \in [-1, 1]$ and $\phi \in [0, 2\pi)$, $z = 0$ are along the cylinder and around the cylinder, respectively.

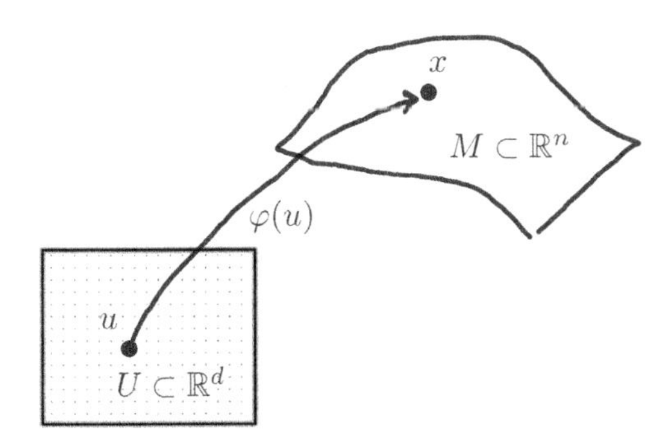

Figure 4.5: Coordinate representation of an arbitrary manifold M using the mapping $\varphi : U \mapsto M$.

Definition 4.2.1 A chart φ is a differentiable and invertible mapping from $U \subset \mathbb{R}^d \mapsto \mathbb{R}^n$, such that $u \in U$ and $\varphi(u) \in M$.

d is the dimension of M. The position vector $x \in \mathbb{R}^n$ is a function of another vector $u \in \mathbb{R}^d$. u^j are the coordinate representations of points in M, playing the same role as θ, ϕ in Exercise 4.1.1. Figure 4.5 illustrates the mapping $\varphi(u)$.

We like to highlight that the convention of defining the chart in the literature is opposite to how we define the chart here. See for example in John Lee [8], the chart is define to map from M to U.

It is sufficient to consider only the open subsets $u \in U \subset \mathbb{R}^d$ because we are often interested in local properties, such as differentials. For global analysis, one considers multiple overlapping coordinate representations. We shall not discuss global geometry in this book.

4.2.1 Scalar functions on M and their differentials

With coordinate representation, we can define the differentials of scalar functions:

$$\begin{aligned} \varphi &: U \mapsto M \\ f &: M \mapsto \mathbb{R} \end{aligned}$$

Then,

$$f \circ \varphi : U \mapsto \mathbb{R}$$

Since $U \subset \mathbb{R}^d$, we are in a position to perform differentiations in the usual calculus manner. Define differentials on f as

$$\frac{\partial f(\varphi(u))}{\partial u^j} = \lim_{\epsilon \to 0} \frac{f(\varphi(u^1, \ldots, u^j + \epsilon, \ldots, u^d)) - f(u^1, \ldots, u^d)}{\epsilon}$$

Note that we cannot take $\partial f / \partial x$ by considering x as a coordinate point in $\mathbb{R}^n$ because f does not exist outside of M.

Special notes on mathematical notations

Throughout the book, we define functions that map between manifolds, e.g. $M \mapsto N$ or $M \mapsto \mathbb{R}^n$. As stated, differentiation cannot be performed on these functions in the embedded space of M. One always uses the coordinate representation by making a composite function:

$$\begin{aligned} f &: M \mapsto \mathbb{R} \\ \varphi &: U \mapsto M \\ f(u) &\equiv f(\varphi(u)) = (f \circ \varphi)(u) \end{aligned}$$

Note that here, we used one symbol, f, to represent both the map $M \mapsto \mathbb{R}$ and the composite map $U \mapsto \mathbb{R}$. There is a one-to-one mapping between f and $f \circ \varphi$. Hence, for shorthand notation, we use the same symbol f to represent it in abstract form and in its coordinate representation form. Usually, the context will make it clear what f refers to. Indeed, when differentials with respect to coordinates are given, then f must be written in coordinate representation.

Example

We give an example to illustrate, $u = \theta$ and $x = (x^1, x^2)$:

$$\begin{aligned} x^1 &= \cos\theta \\ x^2 &= \sin\theta \\ f(x^1, x^2) &= x^1 x^2 + 2 \end{aligned}$$

Then, the coordinate representation of f is

$$f(\theta) = \cos\theta \sin\theta + 2$$

When differentiation is given, then f must be in coordinate representation, e.g. $\partial f / \partial \theta$.

4.3 Tangent Vectors

In the previous section, we introduced a 2D sphere in a 3D Euclidean space. Imagine that you shrink to a very small size and you are standing on a point on this sphere. You look around. What would you observe? You will find that you are on a plane. Just like when you are moving around in your city, you see that the land is quite flat. As you walk along, you are really walking on the surface. How do we represent a local plane in a sphere? This local plane is called the tangent plane. Since this plane is 2D, we can build two 2D basis vectors.

While we denote the manifold as M, at every point $x \in M$, there is a tangent plane with tangent basis vectors spanning this plane:

$$\forall x \in M \subset \mathbb{R}^n, \quad \exists \hat{e}_i \in T_x M \subset \mathbb{R}^d, i = 1, \ldots, d \tag{4.3}$$

Note that $x = \varphi(u)$ is the chart mapping U to M.

4.3.1 How to build basis vectors in a tangent space?

One naive way is to do what we did with the 2D Euclidean space: set arbitrary e_1, e_2 and then use them throughout the whole of the 2D space.

Since a surface is also 2D, can't we just do the same? Recall the examples of Lian, Beng, Seng and Huei: Seng is to the north of Lian, etc. Obviously, these four are on the Earth, and the Earth is a 2D surface in a 3D space. We use north–south–east–west in our navigation, and it works, doesn't it? There are good logical reasons why we cannot assign basis vectors like we do in a Euclidean space. From a 3D global perspective, basis vectors cannot be constant over the whole sphere if we want the basis vectors to always be tangent to the surface.

Recall how we define parameterization using polar coordinates:

$$\begin{aligned} x^1 &= r\cos\theta \\ x^2 &= r\sin\theta \end{aligned}$$

Then, we want the basis vectors in the polar coordinates to be the change in position vector due to a small change in parameter: Let $\varphi(r,\theta) = x^1 e_1 + x^2 e_2 = r\cos\theta e_1 + r\sin\theta e_2$

$$\begin{aligned} \hat{e}_1 &= \frac{\partial\varphi}{\partial r} \\ \hat{e}_2 &= \frac{\partial\varphi}{\partial\theta} \end{aligned}$$

We apply the same principle to basis vectors on a 2-sphere. First, we use the coordinate representation:

$$\begin{aligned} x^1(\theta,\phi) &= \cos\phi\sin\theta \\ x^2(\theta,\phi) &= \sin\phi\sin\theta \\ x^3(\theta,\phi) &= \cos\theta \end{aligned}$$

Then, we define basis vectors in the same way as we did for polar coordinates. Let a point be

$$\begin{aligned} \varphi(\theta,\phi) &= x^1 e_1 + x^2 e_2 + x^3 e_3 \\ &= \cos\phi\sin\theta e_1 + \sin\phi\sin\theta e_2 + \cos\theta e_3 \end{aligned}$$

$$\begin{aligned} \hat{e}_1 &= \frac{\partial x(\theta,\phi)}{\partial\theta} \\ &= \cos\phi\cos\theta e_1 + \sin\phi\cos\theta e_2 - \sin\theta e_3 \\ \hat{e}_2 &= \frac{\partial x(\theta,\phi)}{\partial\phi} \\ &= -\sin\phi\sin\theta e_1 + \cos\phi\sin\theta e_2 \end{aligned} \tag{4.4}$$

Now, we get $\hat{e}_1, \hat{e}_2$; are they correct? The above equations look like a bunch of symbols. How can we check? We can see that the Jacobian is

$$J = \begin{pmatrix} \cos\phi\cos\theta & -\sin\phi\sin\theta \\ \sin\phi\cos\theta & \cos\phi\sin\theta \\ -\sin\theta & 0 \end{pmatrix}$$

Remarks

Note that each value of θ, ϕ corresponds to one point in a 3D space, and all these points put together form a sphere. We are tempted to write position vectors as

$$x = \theta\hat{e}_1 + \phi\hat{e}_2 \tag{4.5}$$

However, this is incorrect. For position vectors, one starts from the origin and then moves in the directions of the basis vectors in sequence until we get to the point we want to. In the case of the 2-sphere, the origin itself is not on the sphere. We cannot "get there." Hence, $\hat{e}_1, \hat{e}_2$ are not the basis vectors of positions. They form the tangent basis at the point x. An important point to note is that e_1, e_2, e_3 are not tangent vectors. Also, for arbitrary coefficients, $x^1 e_1 + x^2 e_2 + x^3 e_3$ may not lie on the unit sphere.

4.3.2 Tangent vector fields

Now, we are ready to give the general definition of tangent vectors and tangent vector fields:

$$\begin{aligned} \varphi(u) &= \varphi^k(u)e_k \\ J^k{}_i &= \frac{\partial\varphi^k(u)}{\partial u^i} \end{aligned}$$

$$\hat{e}_i = J^k{}_i e_k, \qquad i = 1, \ldots, d \text{ and } k = 1, \ldots, n$$

$$\hat{e}_i = \frac{\partial\varphi(u)}{\partial u^i}$$

Figure 4.6 illustrates how the tangent vector can be constructed by infinitesimal increments of the coordinate representation along u^i and u^j. These infinitesimal increments map via φ to infinitesimal changes in the image of φ. The differential of the change in φ w.r.t. Δu gives the tangent vectors.

Definition 4.3.1 A tangent space $T_x M = \mathbb{R}^d$ at the point $x \in M$ is spanned by basis vectors:

$$\hat{e}_i = \frac{\partial\varphi(u)}{\partial u^i}, \qquad i = 1, \ldots, d$$

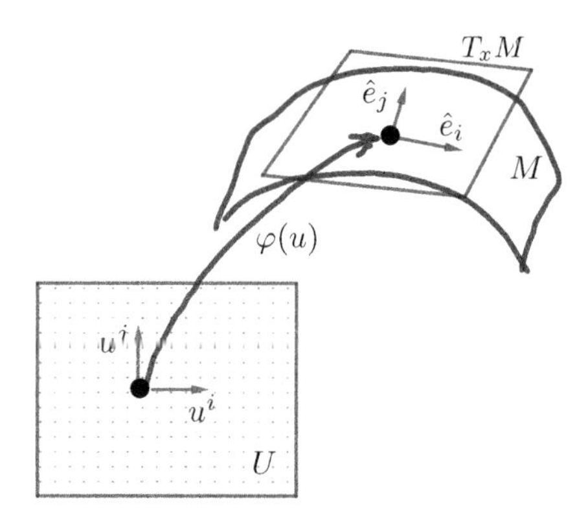

Figure 4.6: Illustration of the chart $\varphi : U \mapsto M$ and tangent vectors generated by infinitesimal increments of the coordinate representation $u + \Delta u$, which translates into infinitesimal shifts in the points on M, $\varphi(u) \to \varphi(u + \Delta u)$. Tangent vectors are $e_i = \partial\varphi/\partial u^i$.

Definition 4.3.1 specifies the basis tangent vector at a point, with the derivative evaluated at the point. In general, the tangent basis vector is not a constant over M. A tangent vector field specifies tangent vectors at each point of M. While we denote position vectors in lower case (e.g. x), we denote tangent vectors in upper case X.

Definition 4.3.2 (differentiable tangent vector field) The tangent vectors at each point of M are given by

$$X = X^i(u)\hat{e}_i = X^i(u)\frac{\partial\varphi}{\partial u^i}$$

Its components, $X^i : U \mapsto \mathbb{R}$, are differentiable functions.

Exercise 4.3.1

1. Verify that the vectors derived in Eq. (4.4) are tangential to the surface of the sphere.

   Solution: We know that these vectors must lie on the surface of the sphere. Therefore, they must be perpendicular to a vector

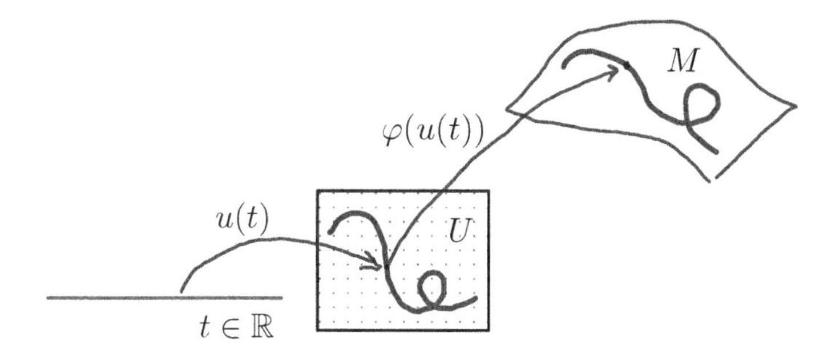

Figure 4.7: Parameterization of a path in M.

shooting from the center of the sphere:

$$\begin{aligned}
\hat{e}_1 \cdot x &= (\cos\phi\cos\theta e_1 + \sin\phi\cos\theta e_2 - \sin\theta e_3)\cdot \\
&\quad (\cos\phi\sin\theta e_1 + \sin\phi\sin\theta e_2 + \cos\theta e_3) \\
&= \cos^2\phi\cos\theta\sin\theta + \sin^2\phi\cos\theta\sin\theta - \sin\theta\cos\theta = 0 \\
\hat{e}_2 \cdot x &= (-\sin\phi\sin\theta e_1 + \cos\phi\sin\theta e_2)\cdot \\
&\quad (\cos\phi\sin\theta e_1 + \sin\phi\sin\theta e_2 + \cos\theta e_3) \\
&= -\sin\phi\cos\phi\sin^2\theta + \cos\phi\sin\phi\sin^2\theta = 0
\end{aligned}$$

Are these two basis vectors perpendicular?

$$\begin{aligned}
e_1 \cdot e_2 &= (\cos\phi\cos\theta e_1 + \sin\phi\cos\theta e_2 - \sin\theta e_3)\cdot \\
&\quad (-\sin\phi\sin\theta e_1 + \cos\phi\sin\theta e_2) \\
&= -\cos\phi\sin\phi\cos\theta\sin\theta + \sin\phi\cos\phi\cos\theta\sin\theta = 0
\end{aligned}$$

4.3.3 Tangent vectors and parametric path

Tangent vectors can be written as the tangential direction of a parametric path in curved space. For example, we can define a path on the 2-sphere by parameterizing the coordinate representations by $t \in \mathbb{R}$, e.g.

$$\begin{aligned}
\theta &= \theta(t) \\
\phi &= \phi(t)
\end{aligned}$$

We can define a parametric path in a general manner. Let u^i be a coordinate representation of a curved space. Then, we define a path on this space as

$$u^i = u^i(t), i = 1, \ldots, d \text{ or } u = u(t)$$

Figure 4.7 illustrates the mapping from $\mathbb{R}$ to a path in M. We can then ask the question, "What is the direction of the tangent vector associated with this path in the curved space?"

1. $u(t)$ defines a point $\varphi(u(t))$ in the curved space.
2. By increasing t infinitesimally, we reach another point x' given by $\varphi(u(t+dt))$.
3. What is the direction of $x' - x$?

Let $\gamma(t) = \varphi(u(t))$ be the path in M:

$$\dot{\gamma}(t) = \lim_{\epsilon \to 0} \frac{\varphi(u(t+\epsilon)) - \varphi(u(t))}{\epsilon}$$

Expanding to first-order ϵ,

$$u^i(t+\epsilon) = u^i(t) + \left.\frac{du^i}{dt}\right|_t \epsilon + O(\epsilon^2)$$

Substituting,

$$\begin{aligned}
\dot{\gamma}(t) &= \lim_{\epsilon \to 0} \frac{\varphi(u(t)) + \frac{du(t)}{dt}\epsilon + O(\epsilon^2) - \varphi(u(t))}{\epsilon} \qquad (4.6)\\
&= \lim_{\epsilon \to 0} \frac{\varphi(u(t)) + \frac{\partial \varphi}{\partial u^i}\frac{du^i(t)}{dt}\epsilon + O(\epsilon^2) - \varphi(u(t))}{\epsilon} \\
\dot{\gamma}(t) &= \frac{\partial \varphi}{\partial u^i}\frac{du^i}{dt} = \sum_i \frac{\partial \varphi}{\partial u^i}\frac{du^i}{dt} \\
&= \frac{du^i}{dt}\hat{e}_i = \dot{\gamma}^i \hat{e}_i
\end{aligned}$$

The summation is explicitly shown to remind readers about Einstein's notation of summing over double indices. The components of $\dot{\gamma}$ are

$$\dot{\gamma}^i = \frac{du^i}{dt}$$

Definition 4.3.3 At every point on a path $\gamma(t)$ in M, there is a corresponding tangent vector $X(t)$:

$$X(t) = X^i \hat{e}_i = \dot{\gamma}(t) = \dot{\gamma}^i \hat{e}_i$$

Exercise 4.3.2

1. Draw the parametric path on the 2-sphere with the functions

$$
\begin{aligned}
\theta(t) &= \sin(t) + 0.1 \\
\phi(t) &= t \\
t &\in [0, \pi]
\end{aligned}
$$

Solution: The path is given by $\theta(t) = \sin(\phi(t)) + 0.1$, with $\theta(0) = 0.1, \phi(0) = 0$. The reader can use a software for parametric plotting in 3D to visualize this path.

---

2. Draw the parametric path of a cylinder with functions

$$
\begin{aligned}
x^1 &= \cos\phi \\
x^2 &= \sin\phi \\
x^3 &= z
\end{aligned}
$$

$$
\begin{aligned}
z(t) &= \sin(t) \\
\phi(t) &= t \\
t &\in [0, \pi]
\end{aligned}
$$

Solution: The path is given by $z = \sin(\phi)$. The reader can use a software for parametric plotting in 3D to visualize this path.

4.4 Induced Metric Tensor for $M \in \mathbb{R}^n$

If we begin with the metric tensor in Euclidean space, we can compute the induced metric tensor in M using the inner products of tangent basis vectors. Now, we can compute the metric tensor using the embedded space,

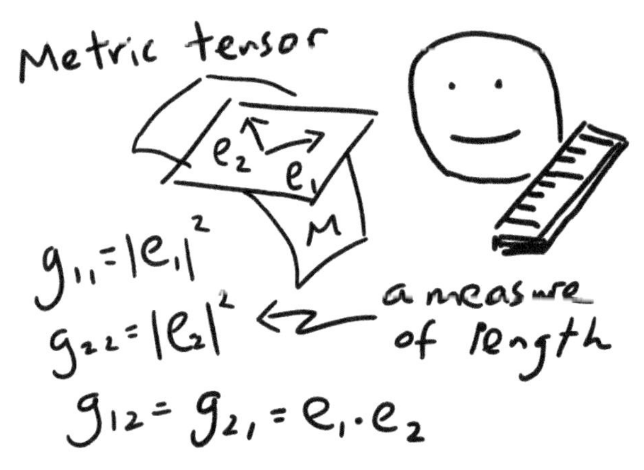

Figure 4.8: The metric tensor induced from the Euclidean metric onto the metric tensor of coordinate representations.

using the 2-sphere as an example:

$$\begin{aligned}
\hat{e}_1 \cdot \hat{e}_1 &= (\cos\phi\cos\theta e_1 + \sin\phi\cos\theta e_2 - \sin\theta e_3)\cdot \\
&\quad (\cos\phi\cos\theta e_1 + \sin\phi\cos\theta e_2 - \sin\theta e_3) \\
&= 1 \\
\hat{e}_1 \cdot \hat{e}_2 &= (\cos\phi\cos\theta e_1 + \sin\phi\cos\theta e_2 - \sin\theta e_3)\cdot \\
&\quad (-\sin\phi\sin\theta e_1 + \cos\phi\sin\theta e_2) = 0 \\
\hat{e}_2 \cdot \hat{e}_1 &= \hat{e}_1 \cdot \hat{e}_2 = 0 \\
\hat{e}_2 \cdot \hat{e}_2 &= (-\sin\phi\sin\theta e_1 + \cos\phi\sin\theta e_2)\cdots(-\sin\phi\sin\theta e_1 + \cos\phi\sin\theta e_2) \\
&= \sin^2\theta
\end{aligned}$$

The metric tensor for a 2-sphere is

$$g = \begin{pmatrix} 1 & 0 \\ 0 & \sin^2\theta \end{pmatrix}$$

Note that $\hat{e}_2$ becomes shorter near the pole and eventually vanishes at the pole. $\hat{e}_1$ maintains its length throughout the unit sphere. The general formula for the induced metric tensor from the Euclidean metric tensor is

$$g_{ij} = \frac{\partial\varphi^k}{\partial u^i}\delta_{ks}\frac{\partial\varphi^s}{\partial u^j}$$

We call this the "pullback" of the Euclidean metric. The word and concept of pullback are explained in later chapters of this book. Figure 4.8 illustrates how two tangent vectors on M can be used to calculate they induced metric tensor.

4.4.1 Calculating the path length of equator in three ways

The metric tensor is used to compute path lengths. We provide an example for computing path lengths for the 2-sphere. The length of the equator for a sphere of unit radius is obviously 2π. This is the simplest and the first way. The second way is to use the arc length formula:

$$l = \int_0^1 \sqrt{\dot{\gamma} \cdot \dot{\gamma}} dt$$

Consider

$$\begin{aligned} \phi(t) &= 2\pi t \\ \theta(t) &= \pi/2 \end{aligned}$$

Then,

$$\begin{aligned} x^1(t) &= \cos(2\pi t)\sin(\pi/2) \\ x^2(t) &= \sin(2\pi t)\sin(\pi/2) \\ x^3(t) &= \cos(\pi/2) \end{aligned}$$

The derivatives are

$$\begin{aligned} \dot{\gamma}^1(t) &= -2\pi \sin(2\pi t) \\ \dot{\gamma}^2(t) &= 2\pi \cos(2\pi t) \\ \dot{\gamma}^3(t) &= 0 \end{aligned}$$

$$\dot{\gamma} \cdot \dot{\gamma} = 4\pi^2 \sin^2(2\pi t) + 4\pi^2 \cos^2(2\pi t) = 4\pi^2$$

The curve length is

$$l = \int_0^1 \sqrt{4\pi^2} dt = 2\pi$$

The third way is to use the 2D representation directly:

$$\begin{aligned} \phi(t) &= 2\pi t \\ \theta(t) &= \pi/2 \end{aligned}$$

$$\begin{aligned} \frac{d\phi}{dt} &= 2\pi \\ \frac{d\theta}{dt} &= 0 \end{aligned}$$

The curve length is

$$l = \int_0^1 \sqrt{\left(\frac{d\theta}{dt}, \frac{d\phi}{dt}\right) \cdot \left(\frac{d\theta}{dt}, \frac{d\phi}{dt}\right)} dt$$

Here, we use the metric tensor for the θ, ϕ basis:

$$\begin{aligned} l &= \int_0^1 \sqrt{\left(\frac{d\theta}{dt}\frac{d\theta}{dt} + \sin^2\theta \frac{d\phi}{dt}\frac{d\phi}{dt}\right)} dt \\ &= \int_0^1 2\pi dt = 2\pi \end{aligned}$$

Among the three ways, the first is the simplest. It seems that we are quite stupid by using cumbersome ways to "go one big round" to do a simple thing. However, the first way of calculating the path length is only applicable for very simple paths on the sphere.

Consider the following path on the sphere and calculate the path length:

$$\begin{aligned} \theta(t) &= t^2 + 0.1 \\ \phi(t) &= t \\ t &\in [0, 1] \end{aligned}$$

$$\begin{aligned} x^1(t) &= \cos(t)\sin(t^2 + 0.1) \\ x^2(t) &= \sin(t)\sin(t^2 + 0.1) \\ x^3(t) &= \cos(t^2 + 0.1) \end{aligned}$$

The derivatives are

$$\begin{aligned} \dot{\gamma}^1(t) &= 2t\cos(t)\cos(t^2 + 0.1) - \sin(t)\sin(t^2 + 0.1) \\ \dot{\gamma}^2(t) &= 2t\sin(t)\cos(t^2 + 0.1) + \sin(t^2 + 0.1)\cos(t) \\ \dot{\gamma}^3(t) &= -2t\sin(t^2 + 0.1) \end{aligned}$$

The above expression is already long. To use the x^1, x^2, x^3 representation, we need to compute $\sqrt{\dot{\gamma}^1\dot{\gamma}^1 + \dot{\gamma}^2\dot{\gamma}^2 + \dot{\gamma}^3\dot{\gamma}^3}$.

Let's consider using the coordinate representation form:

$$\begin{aligned} \frac{d\theta}{dt} &= 2t \\ \frac{d\phi}{dt} &= 1 \end{aligned}$$

Using the metric tensor,

$$\left(\frac{d\theta}{dt}, \frac{d\phi}{dt}\right) \cdot \left(\frac{d\theta}{dt}, \frac{d\phi}{dt}\right) = 4t^2 + \sin^2(t^2 + 0.1)$$

The curve length is

$$l = \int_0^1 \sqrt{4t^2 + \sin^2(t^2 + 0.1)}dt = 1.07937\ldots$$

4.5 The Christoffel Symbols in Curved Space

Equation (3.17) gave the Christoffel symbols in flat space:

$$\frac{\partial e_i}{\partial x^j} = \Gamma^k{}_{ij} e_k \tag{4.7}$$

Now, we need to derive the Christoffel symbols in curved space. Let's consider the unit circle in $\mathbb{R}^2$ as an illustration:

$$\begin{aligned} \varphi^1 = x^1 &= \cos 2\pi u \\ \varphi^2 = x^2 &= \sin 2\pi u \qquad u \in [0, 1] \end{aligned}$$

The tangent vector is

$$\begin{aligned} \hat{e} &= \left(\frac{\partial \varphi^1}{\partial u}, \frac{\partial \varphi^2}{\partial u}\right) \\ &= 2\pi(-\sin 2\pi u, \cos 2\pi u) \end{aligned}$$

Calculate the change of direction in $\hat{e}$:

$$\frac{\partial \hat{e}}{\partial u} = 4\pi^2(-\cos 2\pi u, -\sin 2\pi u) = -4\pi^2 \hat{n} \tag{4.8}$$

$\hat{n}$ is the vector pointing in the radial direction from the origin.

We see that if we use Eq. (4.7) and substitute $e \to \hat{e}$, we observe something strange:

$$\frac{\partial \hat{e}}{\partial u} = \Gamma^1{}_{11} \hat{e}$$

The above equation is wrong because it states that the derivative of $\hat{e}$ is proportional to $\hat{e}$, which is obviously not true in Eq. (4.8). We modify the above equation to make a complete basis in $\mathbb{R}^n$. Add in the normal component of the tangent space:

$$\frac{\partial \hat{e}}{\partial u} = \Gamma^1{}_{11} \hat{e} + L^1{}_{11} \hat{n}$$

Using Eq. (4.8), we see that $\Gamma^1{}_{11} = 0$ and $L^1{}_{11} = -4\pi^2$.

4.5.1 General equation for Christoffel symbols

The general equation is

$$\frac{\partial \hat{e}_i}{\partial u^j} = \Gamma^k{}_{ij}\hat{e}_k + L^k{}_{ij}\hat{n}_k \tag{4.9}$$

$L^k{}_{ij}$ is called the second fundamental form.

Exercise 4.5.1

1. Work out the explicit expression of $\partial \hat{e}_i/\partial u^j$ and the Christoffel symbols for the 2-sphere.

   Solution: Note that there is only one normal vector, which we denote as $\hat{n}$. This vector points radially outward from the origin:

$$\begin{aligned}
\hat{e}_1 &= \cos\phi\cos\theta e_1 + \sin\phi\cos\theta e_2 - \sin\theta e_3 \\
\hat{e}_2 &= -\sin\phi\sin\theta e_1 + \cos\phi\sin\theta e_2 \\
\hat{n} &= \cos\phi\sin\theta e_1 + \sin\phi\sin\theta e_2 + \cos\theta e_3
\end{aligned}$$

$$\begin{aligned}
\frac{\partial \hat{e}_1}{\partial \theta} &= -\cos\phi\sin\theta e_1 - \sin\phi\sin\theta e_2 - \cos\theta e_3 \\
\frac{\partial \hat{e}_1}{\partial \phi} &= -\sin\phi\cos\theta e_1 + \cos\phi\cos\theta e_2 \\
\frac{\partial \hat{e}_2}{\partial \theta} &= -\sin\phi\cos\theta e_1 + \cos\phi\cos\theta e_2 \\
\frac{\partial \hat{e}_2}{\partial \phi} &= -\cos\phi\sin\theta e_1 - \sin\phi\sin\theta e_2
\end{aligned}$$

   Use the metric tensor for $\mathbb{R}^n$ to find the tangential and normal components of the above vectors. For normal components,

$$\begin{aligned}
\frac{\partial \hat{e}_1}{\partial \theta}\cdot\hat{n} &= L_{11} = -1 \\
\frac{\partial \hat{e}_1}{\partial \phi}\cdot\hat{n} &= L_{12} = 0 \\
\frac{\partial \hat{e}_2}{\partial \theta}\cdot\hat{n} &= L_{21} = 0 \\
\frac{\partial \hat{e}_2}{\partial \phi}\cdot\hat{n} &= L_{22} = -\sin^2\theta
\end{aligned}$$

Solve for ${\Gamma^k}_{11}, k = \{1, 2\}$,

$$\begin{aligned}
\frac{\partial \hat{e}_1}{\partial \theta} \cdot \hat{e}_1 &= 0 = {\Gamma^1}_{11}\hat{e}_1 \cdot \hat{e}_1 + {\Gamma^2}_{11}\hat{e}_2 \cdot \hat{e}_1 \\
\frac{\partial \hat{e}_1}{\partial \theta} \cdot \hat{e}_2 &= 0 = {\Gamma^1}_{11}\hat{e}_1 \cdot \hat{e}_2 + {\Gamma^2}_{11}\hat{e}_2 \cdot \hat{e}_2
\end{aligned}$$

We get

$$\begin{pmatrix} 1 & 0 \\ 0 & \sin^2\theta \end{pmatrix} \begin{pmatrix} {\Gamma^1}_{11} \\ {\Gamma^2}_{11} \end{pmatrix} = \begin{pmatrix} 0 \\ 0 \end{pmatrix}$$

$${\Gamma^1}_{11} = {\Gamma^2}_{11} = 0$$

Solve for ${\Gamma^k}_{12}, k = \{1, 2\}$:

$$\begin{aligned}
\frac{\partial \hat{e}_1}{\partial \phi} \cdot \hat{e}_1 &= 0 = {\Gamma^1}_{12}\hat{e}_1 \cdot \hat{e}_1 + {\Gamma^2}_{12}\hat{e}_2 \cdot \hat{e}_1 \\
\frac{\partial \hat{e}_1}{\partial \phi} \cdot \hat{e}_2 &= \cos\theta\sin\theta = {\Gamma^1}_{12}\hat{e}_1 \cdot \hat{e}_2 + {\Gamma^2}_{12}\hat{e}_2 \cdot \hat{e}_2
\end{aligned}$$

We get

$$\begin{pmatrix} 1 & 0 \\ 0 & \sin^2\theta \end{pmatrix} \begin{pmatrix} {\Gamma^1}_{12} \\ {\Gamma^2}_{12} \end{pmatrix} = \begin{pmatrix} 0 \\ \cos\theta\sin\theta \end{pmatrix}$$

$$\begin{aligned}
{\Gamma^1}_{12} &= 0 \\
{\Gamma^2}_{12} &= \cos\theta/\sin\theta
\end{aligned}$$

The Christoffel symbol is symmetric with respect to the two lower indices (why?); hence,

$$\begin{aligned}
{\Gamma^1}_{21} &= 0 \\
{\Gamma^2}_{21} &= \cos\theta/\sin\theta
\end{aligned}$$

Solve for ${\Gamma^k}_{22}, k = \{1, 2\}$:

$$\begin{aligned}
\frac{\partial \hat{e}_2}{\partial \phi} \cdot \hat{e}_1 &= -\cos\theta\sin\theta = {\Gamma^1}_{22}\hat{e}_1 \cdot \hat{e}_1 + {\Gamma^2}_{22}\hat{e}_2 \cdot \hat{e}_1 \\
\frac{\partial \hat{e}_2}{\partial \phi} \cdot \hat{e}_2 &= 0 = {\Gamma^1}_{22}\hat{e}_1 \cdot \hat{e}_2 + {\Gamma^2}_{22}\hat{e}_2 \cdot \hat{e}_2
\end{aligned}$$

We get

$$\begin{pmatrix} 1 & 0 \\ 0 & \sin^2\theta \end{pmatrix} \begin{pmatrix} {\Gamma^1}_{22} \\ {\Gamma^2}_{22} \end{pmatrix} = \begin{pmatrix} -\cos\theta\sin\theta \\ 0 \end{pmatrix}$$

$$\begin{aligned}
\Gamma^1{}_{22} &= -\cos\theta\sin\theta \\
\Gamma^2{}_{22} &= 0
\end{aligned}$$

To summarize,

$$\begin{aligned}
\Gamma^1{}_{11} &= 0 \\
\Gamma^2{}_{11} &= 0 \\
\Gamma^1{}_{12} &= 0 \\
\Gamma^2{}_{12} &= \cos\theta/\sin\theta \\
\Gamma^1{}_{21} &= 0 \\
\Gamma^2{}_{21} &= \cos\theta/\sin\theta \\
\Gamma^1{}_{22} &= -\cos\theta\sin\theta \\
\Gamma^2{}_{22} &= 0
\end{aligned} \tag{4.10}$$

4.6 Geodesics

Geodesics refer to the shortest paths between pairs of points on a manifold. In Euclidean space, the shortest path between two points is the straight line connecting the two points. In a curved space, we try to move between two points as "straight" as possible. Indeed, there is no straight path in a curved manifold. The path has to follow the curvature of the manifold. Figure 4.9 illustrates a curved manifold with two paths. The path shown by the dotted line takes "unnecessary" turns, while the other path shown by the solid line tries as much as possible not to turn. It closely follows the curvature of the manifold. The shortest path between two points on a manifold is the path that does not take turns in the tangential direction. Formally, we want the change of tangent vectors generated along the path to point in the normal direction of the manifold.

4.6.1 Shortest path by the "no turning" method

Let $\gamma(t) = \varphi(u(t))$ trace a path in the embedded space. The basis tangent vectors along this path are

$$\hat{e}_i = \frac{\partial \varphi(u)}{\partial u^i}$$

The second derivative of x is

$$\frac{d\gamma(t)}{dt} = \hat{e}_i \frac{du^i}{dt}$$

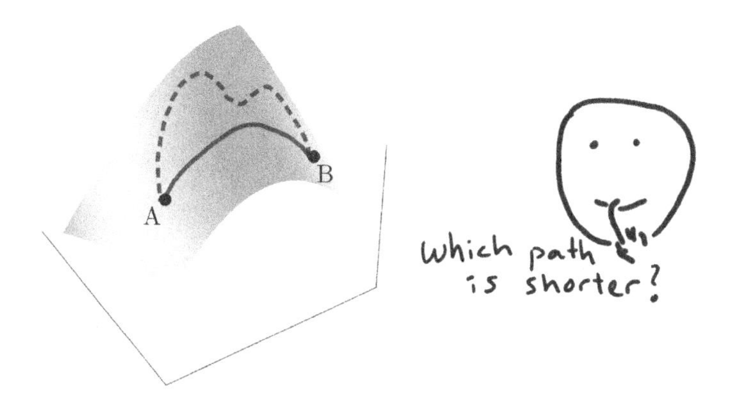

Figure 4.9: Two paths are shown to connect two points A and B on a manifold. The solid line tries as much as possible to follow the curvature of the manifold, while the dotted line takes many unnecessary turns. Hence, the solid line has a shorter length.

$$\begin{aligned}\frac{d^2\gamma}{dt^2} &= \frac{d}{dt}\left(\hat{e}_i\frac{du^i}{dt}\right) \\ &= \frac{d\hat{e}_i}{dt}\frac{du^i}{dt} + \hat{e}_i\frac{d^2u^i}{dt^2} \\ &= \frac{\partial\hat{e}_i}{\partial u^j}\frac{du^j}{dt}\frac{du^i}{dt} + \hat{e}_i\frac{d^2u^i}{dt^2}\end{aligned} \tag{4.11}$$

The second derivative vector can be expressed as a sum of tangent vectors and the normal vector (assume the case of a unique normal vector):

$$\frac{\partial\hat{e}_i}{\partial u^j} = \Gamma^k{}_{ij}\hat{e}_k + L_{ij}\hat{n}$$

For the case that the normal space is more than one dimensional, denote the basis vectors for the normal space as $\hat{n}_k$. Then,

$$\frac{\partial\hat{e}_i}{\partial u^j} = \Gamma^k{}_{ij}\hat{e}_k + L^k{}_{ij}\hat{n}_k \tag{4.12}$$

L_{ij} is called the second fundamental form, and $\Gamma^k{}_{ij}$ are called the Christoffel symbols, as defined in Eq. (3.18). Note that all bases are expressed in the embedded space. Substitute the equation for the second derivative back into the acceleration equation, Eq. (4.11):

$$\begin{aligned}\frac{d^2\gamma}{dt^2} &= \frac{\partial \hat{e}_i}{\partial u^j}\frac{du^j}{dt}\frac{du^i}{dt} + \hat{e}_i\frac{d^2u^i}{dt^2} \\ &= \Gamma^k{}_{ij}\hat{e}_k\frac{du^j}{dt}\frac{du^i}{dt} + \frac{d^2u^k}{dt^2}\hat{e}_k + L^k{}_{ij}\hat{n}_k\frac{du^j}{dt}\frac{du^i}{dt} \\ &= \left(\Gamma^k{}_{ij}\frac{du^j}{dt}\frac{du^i}{dt} + \frac{d^2u^k}{dt^2}\right)\hat{e}_k + L^k{}_{ij}\hat{n}_k\frac{du^j}{dt}\frac{du^i}{dt}\end{aligned}$$

For "no turning," we want the tangential component of the second derivative to be zero. That means

$$\Gamma^k{}_{ij}\frac{du^j}{dt}\frac{du^i}{dt} + \frac{d^2u^k}{dt^2} = 0 \qquad \text{(Geodesic equation)} \tag{4.13}$$

The above equation is a differential equation. If the parameterized paths satisfy these differential equations, then the path is the shortest path. Conversely, given this equation, we solve this differential equation for u^i, and then we get the shortest path. Furthermore, the expression

$$L^k{}_{ij}\frac{du^j}{dt}\frac{du^i}{dt}$$

tells us the extent of the bending of the curved surface when the path is a geodesic.

Computing $L^k{}_{ij}$

Use Eq. (4.12) and apply the dot product on both sides with $\hat{n}$:

$$\frac{\partial \hat{e}_i}{\partial u^j}\cdot \hat{n}_l = L^k{}_{ij}\hat{n}_k \cdot \hat{n}_l = L^l{}_{ij}$$

Lemma 4.6.1 *A path which is a geodesic is given by the solution of the geodesic differential equation*

$$\frac{d^2u^k}{dt^2} + \Gamma^k{}_{ij}\frac{du^j}{dt}\frac{du^i}{dt} = 0$$

Exercise 4.6.1

1. What is the shortest path between two points in a 2D Euclidean space?

Solution: The Christoffel symbols are zeros in a Euclidean space. This can be verified simply by considering that the "turning" of the basis vectors from one point to a neighborhood is zero. So, the geodesic equation becomes

$$\frac{d^2u^k(t)}{dt^2} = 0$$

which yields the solution to a straight line.

---

2. What is the shortest path between two points on a 2D sphere?

Solution: The parameters for a 2D sphere are $u^1 = \theta(t)$ and $u^2 = \phi(t)$. Use the geodesic equation and expand out the indices in the θ and ϕ directions, and use the Christoffel symbols for this parameterization. Note that most of the Christoffel symbols are zeros. So, we reduce the geodesic equation to the following equations:

$$\begin{aligned} \frac{d^2\theta}{dt^2} - \frac{d\phi}{dt}\frac{d\phi}{dt}\cos\theta\sin\theta &= 0 \\ \frac{d^2\phi}{dt^2} + 2\frac{d\theta}{dt}\frac{d\phi}{dt}\frac{\cos\theta}{\sin\theta} &= 0 \end{aligned}$$

Write $d\theta/dt = \dot{\theta}$ and $d\phi/dt = \dot{\phi}$, and the above equation becomes

$$\begin{aligned} \ddot{\theta} - \dot{\phi}^2\cos\theta\sin\theta &= 0 \\ \ddot{\phi} + 2\dot{\theta}\dot{\phi}\frac{\cos\theta}{\sin\theta} &= 0 \end{aligned} \tag{4.14}$$

Then,[1] multiply the first equation by $2\dot{\theta}$ and the second equation by $2\dot{\phi}\sin^2\theta$:

$$\frac{d}{dt}\left(\dot{\theta}^2 + \dot{\phi}^2\sin^2\theta\right) = 0 \tag{4.15}$$

The above equation is significant because it is just the infinitesimal arc length. Recall using the metric tensor for the sphere to get the arc length:

$$\begin{aligned} ds^2 &= \begin{pmatrix} d\theta & d\phi \end{pmatrix} \begin{pmatrix} 1 & 0 \\ 0 & \sin^2\theta \end{pmatrix} \begin{pmatrix} d\theta \\ d\phi \end{pmatrix} \\ &= d\theta^2 + \sin^2\theta d\phi^2 \end{aligned}$$

[1] https://math.stackexchange.com/questions/2723793/geodesics-of-the-unit-sphere-using-christoffel-symbols.

For any initial conditions, $\dot{\theta}(0), \theta(0), \dot{\phi}(0), \phi(0)$, by symmetry of the sphere, we can always rotate the sphere in such a way that in the new coordinate system, $\dot{\theta}(0) = 0$, $\theta(0) = \pi/2$, $\phi(0) = 0$. Set $\dot{\phi}(0) = 1$. Substitute these conditions into Eq. (4.15). We obtain

$$\dot{\theta}^2 + \dot{\phi}^2 \sin^2 \theta = 1 \tag{4.16}$$

Using the above equation to eliminate $\dot{\phi}$ in Eq. (4.14), we get

$$\ddot{\theta} = (1 - \dot{\theta}^2) \cot \theta$$

Rewrite the second-order derivative as

$$\ddot{\theta} = \frac{d\dot{\theta}}{dt} = \frac{d\theta}{dt}\frac{d\dot{\theta}}{d\theta} = \dot{\theta}\frac{d\dot{\theta}}{d\theta} = \frac{d}{d\theta}\left(\frac{1}{2}\dot{\theta}^2\right) = -\frac{1}{2}\frac{d}{d\theta}(1 - \dot{\theta}^2)$$

Let $\chi = 1 - \dot{\theta}^2$. Then,

$$\frac{d\chi}{\chi} + 2\frac{d \sin \theta}{\sin \theta} = 0$$

$$d \ln(\chi \sin^2 \theta) = 0$$

$$\begin{aligned} (1 - \dot{\theta}^2) \sin^2 \theta &= \text{constant} \qquad (4.17) \\ (1 - \dot{\theta}^2(0)) \sin^2 \theta(0) &= 1 \end{aligned}$$

We start with the assumption that the geodesic must lie outside the poles; otherwise, the parametric equation of the sphere has a singularity. Therefore, $\sin \theta \neq 0$. Hence, $\dot{\theta}(t) = 0$ for all t:

$$\begin{aligned} \theta(t) &= a_\theta t + b_\theta \\ &= \pi/2 \end{aligned}$$

with $\dot{\theta}(0) = 0$, and we get $a_\theta = 0$, $\theta(0) = \pi/2$ and $b_\theta = \pi/2$. Using Eq. (4.16),

$$\dot{\phi}(t) = \pm 1$$

$$\phi(t) = a_\phi t + b_\phi$$

with $\dot{\phi}(0) = 1$ and $\phi(0) = 0$, $a_\phi = 1$ and $b_\phi = 0$. The final solution is

$$\begin{aligned} \theta(t) &= \frac{\pi}{2} \\ \phi(t) &= t \end{aligned}$$

This is just the equation of the equator.

---

3. Consider the solution of the geodesic equation on a Euclidean space as $u^*(t)$. With $u^*(0) = \gamma(0), \dot{u}^*(0) = \dot{\gamma}(0)$. Reparameterize the path by $u^*(\lambda t)$: Is this also a solution of the geodesic equation under the same initial condition? If not, what differential equation does $u^*(\lambda t)$ satisfy?

   Solution: First, we find the relationship between the differentials of $u(t)$ and $u(\lambda t)$:

$$\begin{aligned} \frac{du(\lambda t)}{dt} &= \lambda \frac{du(\lambda t)}{d\lambda t} \\ \dot{u}(\lambda t) &= \lambda \dot{u}(t) \\ \ddot{u}(\lambda t) &= \lambda^2 \ddot{u}(t) \end{aligned}$$

   For a Euclidean space,

$$\ddot{u}^*(t) = 0$$

   Therefore,

$$\ddot{u}^*(\lambda t) = 0$$

   with $u^*(t)|_{t=0)} = u^*(\lambda t)|_{t=0}$ and $\dot{u}^*(0) = \lambda \dot{u}^*(0)$.

---

4. Compute the geodesics of a 3D Euclidean space from $(\sin \pi/4, 0, \cos \pi/4)$ to $(0, 1, 0)$.

   Solution: This is a problem with boundary conditions. In general, boundary condition problems are harder to solve:

$$\ddot{u}(t) = 0$$

$$u(t) = at + b$$

   Using the boundary conditions and assuming $t = 1$ at the end point,

$$u(t) = (\sin \pi/4, 0, \cos \pi/4) + ((0, 1, 0) - (\sin \pi/4, 0, \cos \pi/4))t$$

---

5. Compute the length of the geodesic of a 2-sphere starting from $(\theta = \pi/4, \phi = 0$ to $(\theta = \pi/2, \phi = \pi/2)$.

   Solution: Solving for this geodesic differential equation as a boundary value problem is difficult. A common sense approach can easily give us the answer in closed form. We will still solve the differential equation as a boundary value problem for practice and then check our numerical answers with the simple closed-form answer.

We first present the simple closed-form solution. The idea is to take the dot product in the ambient Euclidean space of the two position vectors at the end points. We know that the geodesic is the arc sweeping along a great circle. Then, the length of this arc is equal to the angle between the two end points. The start point x_s is at $(\theta, \phi) = (\pi/4, 0)$, and the end point x_e is at $(\theta, \phi) = (\pi/2, \pi/2)$:

$$x_s = \cos\phi \sin\theta e_1 + \sin\phi \sin\theta e_2 + \cos\theta e_3 = \frac{1}{\sqrt{2}} e_1 + \frac{1}{\sqrt{2}} e_3$$

$$x_e = \cos\phi \sin\theta e_1 + \sin\phi \sin\theta e_2 + \cos\theta e_3 = e_2$$

$x_s \cdot x_e = 0$, which implies that the angle between these two position vectors is $\pi/2$. The arc length is equal to the angle subtending the arc multiplied by the radius of the arc (which is 1). Therefore, the geodesic length is $\pi/2$.

Now, we solve this question using a numerical method, which is unnecessarily hard. Starting from the equations,

$$\begin{aligned}
\ddot{\theta} - \dot{\phi}^2 \cos\theta \sin\theta &= 0 \\
\ddot{\phi} + 2\dot{\phi}\dot{\theta}\frac{\cos\theta}{\sin\theta} &= 0 \\
\dot{\theta}^2 + \dot{\phi}^2 \sin^2\theta &= \dot{\theta}(0)^2 + \dot{\phi}(0)^2 \sin^2\theta(0) = q
\end{aligned}$$

Using the above equation to eliminate $\dot{\phi}$,

$$\ddot{\theta} = \frac{\cos\theta}{\sin\theta}(q - \dot{\theta}^2)$$

Using the same trick as we did for Eq. (4.17),

$$(q - \dot{\theta}^2)\sin^2\theta = \text{constant} = k \tag{4.18}$$

It can be shown that

$$\begin{aligned}
q &= \dot{\theta}(0)^2 + \dot{\phi}(0)^2 \sin^2\theta(0) \\
k &= \dot{\phi}(0)^2 \sin^4\theta(0)
\end{aligned}$$

Using the Euler method to integrate Eq. (4.18),

$$\theta(t + \delta) = \theta(t) + \delta\sqrt{q - \frac{k}{\sin^2\theta(t)}}$$

At the same time, update ϕ using

$$\dot{\theta}(t)^2 + \dot{\phi}(t)^2 \sin^2\theta(t) = q$$

$$\phi(t+\delta) = \phi(t) + \delta\sqrt{\frac{q - \dot{\theta}(t)^2}{\sin^2\theta(t)}}$$

Since $\dot{\theta}(0)$ and $\dot{\phi}(0)$ are unknown, we make an initial guess and then integrate $\theta(t)$ to get the numerically computed end point $\hat{\theta}, \hat{\phi}$. Then, minimize a loss function with respect to $\dot{\theta}(0)$ and $\dot{\phi}(0)$ using the pytorch autograd function:

$$\text{Loss} = (\hat{\theta} - \theta(1))^2 + (\hat{\phi} - \phi(1))^2$$

Solving an equation is only the start of the analysis. We should always check and make discoveries from our solution. There are two ways to check. The first is to check that the numerically computed end point for θ and ϕ is correct by computing

$$\begin{aligned} \epsilon_\theta &= \hat{\theta} - \theta(1) \\ \epsilon_\phi &= \hat{\phi} - \phi(1) \end{aligned}$$

The second way is to check with the known correct answer for the path length, which is $\pi/2$. Let the numerically computed length be $\hat{l}$. Then,

$$\epsilon_l = \hat{l} - \pi/2$$

The source code for this computation is given as follows.

```
# ================================================================
import torch
import numpy as np
import matplotlib.pyplot as plt
from mpl_toolkits.mplot3d import Axes3D

# ================================================================
class int_theta_phi:

    # theta0 gives the initial values of theta
    def __init__(self,theta0,phi0,dt,nsteps):
        self.dt = dt                    # Euler scheme
            time step
        self.nsteps = nsteps       # number of steps
            taken to trace our path
```

```
        # make initial theta as a tensor and
            disable gradient for this variable
        self.theta0 = torch.tensor([theta0],
            requires_grad=False)
        self.phi0 = torch.tensor([phi0],
            requires_grad=False)
#===================================================
# do one step euler update
def _one_step(self,theta,phi,k,q,dotthetasq):
        sinsq = self.sinsq(theta)
        lamt = q-k/sinsq
        dottheta = torch.sqrt(lamt)
        lamp = (q - dotthetasq)/sinsq
        dotphi = torch.sqrt(lamp)
        theta_next = theta + self.dt*dottheta
        phi_next = phi + self.dt*dotphi
        return theta_next,phi_next,lamt,lamp
#===================================================
# trace the path -- dottheta0sq and dotphi0sq
    are
# variables in which we need autograd and to
    optimize
def __call__(self,dottheta0sq,dotphi0sq):
        theta = torch.clone(self.theta0).detach()
        phi = torch.clone(self.phi0).detach()
        dotthetasq = dottheta0sq

        k = self.eval_k(dotphi0sq)
        q = self.eval_q(dottheta0sq,dotphi0sq)

        theta_path    = [theta]
        phi_path      = [phi]
        dthetasq_path = [dottheta0sq]
        dphisq_path   = [dotphi0sq]

        for t in range(self.nsteps):
            theta,phi,dotthetasq,dotphisq = self.
                _one_step(theta,phi,k,q,
                dotthetasq)
            theta_path.append(theta)
            phi_path.append(phi)
```

```
            dthetasq_path.append(dotthetasq)
            dphisq_path.append(dotphisq)

        theta_path    = torch.stack(theta_path)
        phi_path      = torch.stack(phi_path)
        dthetasq_path = torch.stack(dthetasq_path
            )
        dphisq_path   = torch.stack(dphisq_path)

        return theta_path,phi_path,dthetasq_path,
            dphisq_path

    # ==================================================
    def sinsq(self,theta):
        return torch.sin(theta)*torch.sin(theta)

    # compute the variable q in torch.tensor
    def eval_q(self,dottheta0sq,dotphi0sq):
        return dottheta0sq + dotphi0sq*self.sinsq
            (self.theta0)

    # compute the variable k in torch.tensor
    def eval_k(self,dotphi0sq):
        return dotphi0sq*self.sinsq(self.theta0)*
            self.sinsq(self.theta0)

# ======================================================
def compute_length(dt,dthetasq,dphisq,theta):
    sinsq = torch.square(torch.sin(theta))
    dl = torch.sqrt(dthetasq + dphisq*sinsq)
    return dt*torch.sum(dl)
# ======================================================
def plot_path(theta,phi):

    npt = theta.detach().numpy()
    npp = phi.detach().numpy()
    x = np.cos(npp)*np.sin(npt)
    y = np.sin(npp)*np.sin(npt)
    z = np.cos(npt)

    fig = plt.figure()
    ax = fig.add_subplot(111,projection='3d')
```

```
    ax.scatter(x,y,z)
    plt.show()
#==================================================================
if __name__=='__main__':

    #==================================
    # optimizing the variable theta
    #==================================
    nsteps = 1000
    dt = 1./nsteps
    ngrads_theta = 20
    lr_theta = 1e-2

    theta_init  = np.pi/4.0
    theta_final = np.pi/2.0
    phi_init    = 0.0
    phi_final   = np.pi/2.0

    dottheta0 = 0.01 # accurate value=0.0072
    dotphi0   = 2.2  # accurate value=2.2206

    theta_phi_obj = int_theta_phi(theta_init,
       phi_init,dt,nsteps)

    # for finding \dot\theta(0)
    for iter in range(ngrads_theta):

        #print('dtheta0sq',dottheta0sq,'dphi0sq',
           dotphi0sq)
        dt0 = torch.tensor([dottheta0],
           requires_grad=True)
        dp0 = torch.tensor([dotphi0],
           requires_grad=True)

        # call the theta object to trace out a
           path
        # returns the end point
        theta_path,phi_path,_,_ = theta_phi_obj(
           dt0*dt0,dp0*dp0)

        theta = theta_path[-1]
        phi   = phi_path[-1]
```

```
        # mse loss on the end points
        loss = torch.square(theta-theta_final)+
            torch.square(phi-phi_final)
        loss.backward()

        # update dottheta0 and dotphi0
        dottheta0 = (dt0.data - lr_theta*dt0.grad
            .data).item()
        dotphi0   = (dp0.data - lr_theta*dp0.grad
            .data).item()

        if iter%10==0:
            print(iter,'dt0',dt0,'dt0.grad ',dt0.
                grad)
            print(iter,'dp0',dp0,'dp0.grad ',dp0.
                grad)
            print(iter,'loss ',loss)

    # get theta and dottheta as a function of t
    dt0sq = torch.tensor([dottheta0*dottheta0],
        requires_grad=True)
    dp0sq = torch.tensor([dotphi0*dotphi0],
        requires_grad=True)
    theta_path,phi_path,dthetasq_path,dphisq_path
        = theta_phi_obj(dt0sq,dp0sq)

    # ================================
    # after getting path, check if end points
        match
    print('checking if end points matches...end
        point errors')
    dtheta = theta_path[-1]-theta_final
    dphi   = phi_path[-1]-phi_final
    print('dtheta',dtheta.item(),'dphi',dphi.item
        ())
    # check length
    correct_length = np.pi/2.0 # using angle
        between two end points
    length = compute_length(dt,dthetasq_path,
        dphisq_path,theta_path).item()
    print('dlength',length-correct_length)
```

```
# print the path to a text file
plot_path(theta_path, phi_path)
```

4.7 Summary

We shall refer the formal definition of manifolds to a wide range of existing literature.

Definition 4.2.1 A chart φ is a differentiable and invertible mapping from $U \subset \mathbb{R}^d \mapsto \mathbb{R}^n$, such that $u \in U$ and $\varphi(u) \in M$.

We like to highlight that the convention of defining the chart in the literature is opposite to how we define the chart here. See for example, in John Lee [7], the chart is defined from map from M to U.

Definition 4.3.1 A tangent space $T_x M = \mathbb{R}^d$ at the point $x \in M$ is spanned by basis vectors:

$$\hat{e}_i = \frac{\partial \varphi(u)}{\partial u^i}, \qquad i = 1, \ldots, d$$

Definition 4.3.2 (differentiable tangent vector field) The tangent vectors at each point of M are given by

$$X = X^i(u)\hat{e}_i = X^i(u)\frac{\partial \varphi}{\partial u^i}$$

Its components, $X^i : U \mapsto \mathbb{R}$, are differentiable functions.

Definition 4.3.3 At every point on a path $\gamma(t)$ in M, there is a corresponding tangent vector $X(t)$:

$$X(t) = X^i \hat{e}_i = \dot{\gamma}(t) = \dot{\gamma}^i \hat{e}_i$$

Definition 4.7.1 The induced metric tensor of $M \subset \mathbb{R}^n$ is given by

$$g_{ij} = \hat{e}_i \cdot \hat{e}_j$$

where the dot product is given by the Euclidean dot product.

4.7.1 Christoffel symbols and second fundamental form

The Christoffel symbols and the second fundamental form measure the rate of change of tangent basis vectors at x with respect to a small change in u:

$$\frac{\partial \hat{e}_i}{\partial u^j} = \Gamma^k{}_{ij} \hat{e}_k + L^k{}_{ij} \hat{n}^u_k \tag{4.19}$$

$\Gamma^k{}_{ij}$ are the Christoffel symbols, and $L^k{}_{ij}$ is the second fundamental form. For a 2-sphere, the coordinate representation is

$$\begin{aligned} x^1 &= \cos\phi \sin\theta \\ x^2 &= \sin\phi \sin\theta \\ x^3 &= \cos\theta \end{aligned}$$

With $u^1 = \phi$ and $u^2 = \theta$, the Christoffel symbols are given by

$$\begin{aligned} \Gamma^1{}_{11} &= 0 \\ \Gamma^2{}_{11} &= 0 \\ \Gamma^1{}_{12} &= 0 \\ \Gamma^2{}_{12} &= \cos\theta / \sin\theta \\ \Gamma^1{}_{21} &= 0 \\ \Gamma^2{}_{21} &= \cos\theta / \sin\theta \\ \Gamma^1{}_{22} &= -\cos\theta \sin\theta \\ \Gamma^2{}_{22} &= 0 \end{aligned} \tag{4.20}$$

Definition 4.7.2 A geodesic is a path with the shortest length defined in Definition (3.5.3) with appropriate initial or boundary conditions,

Lemma 4.6.1 *A path which is a geodesic is given by the solution of the geodesic differential equation*

$$\frac{d^2 u^k}{dt^2} + \Gamma^k{}_{ij} \frac{du^j}{dt} \frac{du^i}{dt} = 0$$

Chapter 5

Covariant Derivatives and Parallel Transport on $M \subset \mathbb{R}^n$

In the previous chapter, we discussed the differentiation of a scalar function on a manifold. In this chapter, we explore the differentiation of tangent vectors on a manifold with respect to the coordinate representations. Such a differentiation is more complicated than the differentiation of a scalar function. This is because tangent vectors have directions. We would like to know how much tangent vectors change due to a small change in coordinates. Let's call this change ΔX. We know that the differences of vectors are vectors, and hence ΔX is also a vector. Indeed, the change in tangent vectors (ΔX) as they move from one point on the manifold to an infinitesimally nearby point oftentimes does not lie in any tangent plane. ΔX can point out of the manifold. A covariant derivative is a way to differentiate tangent vectors such that the differentiation results in another vector that lies in the tangent plane. This is achieved by projections onto the tangent plane. It is also common in the literature to call covariant derivatives "connections." Hence, we like to notify the reader that if they read about connections, it can mean covariant derivatives.

5.1 Covariant Derivatives in $\mathbb{R}^n$

Recall that a vector field is such that all given points in the underlying space are assigned a vector. A differentiable vector field is one such that the vectors from one point to their neighboring point do not "change suddenly.

That means the differentials of vectors with respect to the position exist. Let $X(x) = X(x^i e_i)$ be a vector field, on $\mathbb{R}^n$:

$$X(x) = X^j(x)e_j$$

Writing in indices notation,

$$\frac{\partial X}{\partial x^j} = \frac{\partial X^i}{\partial x^j}e_i + X^i\frac{\partial e_i}{\partial x^j}$$

Recall the Christoffel symbols from Eq. (3.16):

$$\frac{\partial e_i}{\partial x^j} = \Gamma^k{}_{ij}e_k$$

Substituting gives

$$\frac{\partial X}{\partial x^j} = \left(\frac{\partial X^k}{\partial x^j} + v^i\Gamma^k{}_{ij}\right)e_k \tag{5.1}$$

Exercise 5.1.1

1. Derive Eq. (5.1).

   Solution: Given

   $$\frac{\partial X}{\partial x^j} = \frac{\partial X^i}{\partial x^j}e_i + X^i\frac{\partial e_i}{\partial x^j}$$

   substitute the expression for Christoffel symbols:

   $$\frac{\partial e_i}{\partial x^j} = \Gamma^k{}_{ij}e_k$$

   We get

   $$\frac{\partial X}{\partial x^j} = \frac{\partial X^i}{\partial x^j}e_i + X^i\Gamma^k{}_{ij}e_k$$

   Reindexing the first term $i \to k$,

   $$\frac{\partial X}{\partial x^j} = \frac{\partial X^k}{\partial x^j}e_k + X^i\Gamma^k{}_{ij}e_k = \left(\frac{\partial X^k}{\partial x^j} + X^i\Gamma^k{}_{ij}\right)e_k$$

5.2 Covariant Derivative in M

As shown in Eq. (4.19), the derivative of a basis vector is different in $\mathbb{R}^n$ and M. In $\mathbb{R}^n$, there is no normal component. In M, there are extra dimensions outside of M for vectors to point into:

$$\frac{\partial \hat{e}_i}{\partial u^j} = \Gamma^k{}_{ij}\hat{e}_k + L^k{}_{ij}\hat{n}_k$$

Now, take the derivatives of an arbitrary tangent vector field on M, $X = X^i\hat{e}_i$:

$$\begin{aligned} \frac{\partial X}{\partial u^j} &= \frac{\partial X^i}{\partial u^j}\hat{e}_i + X^i\frac{\partial \hat{e}_i}{\partial u^j} \\ &= \left(\frac{\partial X^k}{\partial u^j} + X^i\Gamma^k{}_{ij}\right)\hat{e}_k + X^iL^k{}_{ij}\hat{n}_k \end{aligned} \tag{5.2}$$

We want to construct an operation that, when we take derivatives of vector fields in M, the resulting vector is on T_xM. Indeed, for a curved space, this is impossible with the usual calculus derivative. We need to define the covariant derivative in a slightly different way. Different symbols are used in different publications; we use $D_{\hat{e}_j}$ to represent covariant derivative, where we differentiate in the direction of $\hat{e}_j$. Lastly, to force $D_{\hat{e}_j}$ to generate a vector on T_xM, we project the result of the derivative (Eq. (5.2)) onto T_xM, which means that we throw away all components not tangent to M:

$$D_{\hat{e}_j}X = \left(\frac{\partial X^k}{\partial u^j} + X^i\Gamma^k{}_{ij}\right)\hat{e}_k \in T_xM \tag{5.3}$$

The covariant derivative of X in a general direction $Y = Y^i\hat{e}_i$ is

$$D_YX = Y^iD_{\hat{e}_i}X$$

5.3 Important Properties of Covariant Derivatives

Lemma 5.3.1 *Let D_YX be the covariant derivative of X in the direction $Y \in T_xM$. Let $Z \in T_xM$ and $X \in T_xM$. Then, the following identities hold:*

1. *Linear in the direction of derivative: $f, g : M \mapsto \mathbb{R}$.*

$$D_{fY_1+gZ_2}X = fD_YX + gD_ZX \tag{5.4}$$

2. *Linear in input:* $a_1, a_2 \in \mathbb{R}$.

$$D_Z(a_1 X + a_2 Y) = a_1 D_Z X + a_2 D_Z Y \tag{5.5}$$

3. *Product rule,* $f : M \mapsto \mathbb{R}$:

$$D_Y(fX) = f D_Y X + X D_Y f \tag{5.6}$$

4. *Metric compatibility:*

$$D_Z(X \cdot Y) = (D_Z X) \cdot Y + X \cdot (D_Z Y) \tag{5.7}$$

5. *Torsion-free property:*

$$D_X Y - D_Y X = [X, Y] = \left(X^i \frac{\partial Y^j}{\partial u^i} - Y^i \frac{\partial X^j}{\partial u^i} \right) \hat{e}_j \tag{5.8}$$

5.3.1 Covariant derivative on scalar functions

Given a function,

$$f : M \mapsto \mathbb{R}$$

the covariant derivative of f is

$$D_{\hat{e}_j} f = \frac{\partial f \circ \varphi(u)}{\partial u^j} = \frac{\partial f(u)}{\partial u^j}$$

Here again, we use the same symbols for the function $f : M \mapsto \mathbb{R}$ and the function $f \circ x : U \mapsto \mathbb{R}$, i.e. $f \equiv f \circ x$.

5.3.2 Covariant derivative as an operator

It is common in the literature to call covariant derivatives "connections." The nabla mathematical symbol (∇) is also often used to denote the covariant derivatives. We shall stick to the notation in this book: we use D to represent the covariant derivative. Indeed, the covariant derivative maps two tangent vectors to another tangent vector:

$$D : T_x M \times T_x M \mapsto T_x M$$

Denote by $D_X Y = D(X, Y)$ a bilinear function, for $a, b \in \mathbb{R}$, $f, g : M \mapsto \mathbb{R}$:

$$\begin{aligned} D(fX + gY, Z) &= fD(X, Z) + gD(Y, Z) \\ D(X, aY + bZ) &= aD(X, Y) + bD(X, Z) \end{aligned}$$

5.4 Covariant Derivative on a Path

Some literature define the covariant derivative of a vector field in terms of differentiating along a path. This alternative definition results in the same mathematical expression as that mentioned in the previous sections of this chapter. Consider a path $\gamma(t) \in M \subset \mathbb{R}^n$. The direction of the path is given by

$$\begin{aligned} \frac{d\varphi}{dt} &= \lim_{\epsilon \to 0} \frac{\varphi(u(t+\epsilon)) - \varphi(u(t))}{\epsilon} \\ \dot{\gamma} &= \frac{\partial \varphi}{\partial u^j}\frac{du^j}{dt} = \frac{du^j}{dt}\hat{e}_j \end{aligned}$$

Given a vector field $X = X^i \hat{e}_i$, the covariant derivative along this path is the usual derivative with results projected onto the tangent plane $T_x M$. With $X^i = X^i(u)$,

$$\begin{aligned} \frac{dX^i(u)}{dt} &= \frac{\partial X^i}{\partial u^j}\frac{du^j}{dt} \\ \frac{d\hat{e}_i(u)}{dt} &= \frac{\partial \hat{e}_i}{\partial u^j}\frac{du^j}{dt} \end{aligned}$$

$$\begin{aligned} \frac{dX}{dt} &= \frac{dX^i}{dt}\hat{e}_i + X^i \frac{d\hat{e}_i}{dt} \qquad (5.9) \\ &= \left(\frac{\partial X^i}{\partial u^j}\hat{e}_i + X^i \frac{\partial \hat{e}_i}{\partial u^j} \right) \frac{du^j}{dt} \end{aligned}$$

We get one term in the parenthesis that is identical to the derivative of the vector field discussed in the previous section. Hence, differentiation along the path $\gamma(t)$ with respect to t and projection onto $T_x M$ is given by

$$D_{\dot{\gamma}} X = \left(D_{\hat{e}_j} X\right) \frac{du^j}{dt} = \left(D_{\hat{e}_j} X\right) \dot{\gamma}^j \qquad (5.10)$$

Note that the above equation factorizes into two terms. The first term depends on the input vector X, while the second term depends on the direction of the covariant derivative $\dot{\gamma}$. This is an important property for the linearity of covariant derivatives.

5.5 Christoffel Symbols and Their Transformation

Now, we see how the Christoffel symbols transform under coordinate transformation in $T_x M$. Let the tangent basis vectors in these two coordinates be e_l and $\hat{e}_i$:

$$\begin{aligned} \hat{e}_i &= J^k{}_i e_k \\ X^i e_i &= \hat{X}^i \hat{e}_i \\ \partial_{\hat{u}^i} &= J^j{}_i \partial_{u^j} \\ \hat{X}^i &= J^{-1}{}^i{}_k X^k = B^i{}_k X^k \\ B^i{}_k J^k{}_j = J^{-1}{}^i{}_k J^k{}_j &= \delta^i{}_j \end{aligned} \tag{5.11}$$

For simplicity of notation, let $B = J^{-1}$. The covariant derivatives in both coordinate systems must be the same:

$$D_{\hat{e}_j}(\hat{X}^i \hat{e}_i) = D_{J^l{}_j e_l}(X^i e_i) = J^l{}_j D_{e_l}(X^i e_i) \tag{5.12}$$

$$\left(\frac{\partial \hat{X}^k}{\partial \hat{u}^j} + \hat{X}^i \hat{\Gamma}^k_{ij} \right) \hat{e}^u_k = J^l{}_j \left(\frac{\partial X^t}{\partial u^l} + X^i \Gamma^t{}_{il} \right) e_t \tag{5.13}$$

Substituting all the transformation rules and keeping in mind that transformation rules may depend on the point in M,

$$\left(J^l{}_j \frac{\partial B^k{}_s X^s}{\partial u^l} + B^i{}_s X^s \hat{\Gamma}^k_{ij} \right) J^t{}_k e_t = J^l{}_j \left(\frac{\partial X^t}{\partial u^l} + X^i \Gamma^t{}_{il} \right) e_t \tag{5.14}$$

Considering that the components of each e_t are equal and rearranging the indices,

$$\begin{aligned} J^t{}_k J^l{}_j \frac{\partial B^k{}_s X^s}{\partial u^l} + J^t{}_k B^i{}_s X^s \hat{\Gamma}^k_{ij} &= J^l{}_j \frac{\partial X^t}{\partial u^l} + J^l{}_j X^s \Gamma^t{}_{sl} \\ J^t{}_k J^l{}_j \frac{\partial B^k{}_s}{\partial u^l} X^s + J^l{}_j J^t{}_k B^k{}_s \frac{\partial X^s}{\partial u^l} + J^t{}_k B^i{}_s X^s \hat{\Gamma}^k_{ij} &= J^l{}_j \frac{\partial X^t}{\partial u^l} + J^l{}_j X^s \Gamma^t{}_{sl} \\ J^t{}_k J^l{}_j \frac{\partial B^k{}_s}{\partial u^l} X^s + J^l{}_j \delta^t{}_s \frac{\partial X^s}{\partial u^l} + J^t{}_k B^i{}_s X^s \hat{\Gamma}^k_{ij} &= J^l{}_j \frac{\partial X^t}{\partial u^l} + J^l{}_j X^s \Gamma^t{}_{sl} \\ J^t{}_k J^l{}_j \frac{\partial B^k{}_s}{\partial u^l} X^s + J^l{}_j \frac{\partial X^t}{\partial u^l} + J^t{}_k B^i{}_s X^s \hat{\Gamma}^k_{ij} &= J^l{}_j \frac{\partial X^t}{\partial u^l} + J^l{}_j X^s \Gamma^t{}_{sl} \end{aligned}$$

The second term on the left-hand side cancels with the first term in right-hand side, and factorizing X^s,

$$\left(J^t{}_k J^l{}_j \frac{\partial B^k{}_s}{\partial u^l} + J^t{}_k B^i{}_s \hat{\Gamma}^k_{ij} \right) X^s = J^l{}_j \Gamma^t{}_{sl} X^s \tag{5.15}$$

Since X^s is arbitrary, the factors of X^s must be equal:

$$J^t{}_k J^l{}_j \frac{\partial B^k{}_s}{\partial u^l} + J^t{}_k B^i{}_s \hat{\Gamma}^k_{ij} = J^l{}_j \Gamma^t{}_{sl} \tag{5.16}$$

Rearranging and performing back transformations,

$$\begin{aligned} J^t{}_k B^i{}_s \hat{\Gamma}^k_{ij} &= J^l{}_j \Gamma^t{}_{sl} - J^t{}_k J^l{}_j \frac{\partial B^k{}_s}{\partial u^l} \\ B^m{}_t J^t{}_k B^i{}_s \hat{\Gamma}^k_{ij} &= B^m{}_t J^l{}_j \Gamma^t{}_{sl} - B^m{}_t J^t{}_k J^l{}_j \frac{\partial B^k{}_s}{\partial u^l} \\ \delta^m{}_k B^i{}_s \hat{\Gamma}^k_{ij} &= B^m{}_t J^l{}_j \Gamma^t{}_{sl} - \delta^m{}_k J^l{}_j \frac{\partial B^k{}_s}{\partial u^l} \\ B^i{}_s \hat{\Gamma}^m_{ij} &= B^m{}_t J^l{}_j \Gamma^t{}_{sl} - J^l{}_j \frac{\partial B^m{}_s}{\partial u^l} \\ J^s{}_q B^i{}_s \hat{\Gamma}^m_{ij} &= J^s{}_q B^m{}_t J^l{}_j \Gamma^t{}_{sl} - J^s{}_q J^l{}_j \frac{\partial B^m{}_s}{\partial u^l} \\ \delta^i{}_q \hat{\Gamma}^m_{ij} &= J^s{}_q B^m{}_t J^l{}_j \Gamma^t{}_{sl} - J^s{}_q J^l{}_j \frac{\partial B^m{}_s}{\partial u^l} \\ \hat{\Gamma}^m_{qj} &= J^s{}_q B^m{}_t J^l{}_j \Gamma^t{}_{sl} - J^s{}_q J^l{}_j \frac{\partial B^m{}_s}{\partial u^l} \end{aligned} \tag{5.17}$$

We are doing this for what purpose? The above equation tells us that the Christoffel symbol does not transform like a tensor. However, if $\partial B^m{}_s / \partial u^l = 0$, then the Christoffel symbol follows the tensor transformation law. In what scenario will the Christoffel symbol be constant over the manifold? In Euclidean space, the Christoffel is zero everywhere and is therefore also constant. Does there exist a metric such that the Christoffel symbol is non-zero but constant?

Exercise 5.5.1

1. Prove Eq. (5.4):

$$D_{fY_1 + gZ_2} X = f D_Y X + g D_Z X \tag{5.18}$$

Solution: To prove Eq. (5.4), consider tangent vectors along three paths, $\dot{\gamma}_1, \dot{\gamma}_2$ and $\dot{\gamma} = a_1 \dot{\gamma}_1 + a_2 \dot{\gamma}_2$. The covariant derivatives of a vector along γ_1 and γ_2 are

$$\begin{aligned} D_{\dot{\gamma}_1} X &= \left(D_{\hat{e}_j} X\right) \dot{\gamma}_1^j \\ D_{\dot{\gamma}_2} X &= \left(D_{\hat{e}_j} X\right) \dot{\gamma}_2^j \end{aligned} \tag{5.19}$$

Multiply by scalars a_1, a_2 and adding gives,

$$\left(D_{\hat{e}_j} X\right) \left(a_1 \dot{\gamma}_1^j + a_2 \dot{\gamma}_2^j\right) = \left(D_{\hat{e}_j} X\right) \dot{\gamma}^j = D_{\dot{\gamma}} X \tag{5.20}$$

---

2. Prove Eq. (5.5):

$$D_Z(a_1X + a_2Y) = a_1D_ZX + a_2D_ZY \tag{5.21}$$

Solution: To prove Eq. (5.5), use Eq. (5.4) to reduce the directions into a linear combination of u^j. Then, we need only to prove linearity in the input for one of u^j, and after that, "assemble back." Let $X, Y \in T_xM$. Then,

$$\begin{aligned} D_{\hat{e}_j}(a_1X + a_2Y) &= \left(\frac{\partial a_1X^k + a_2Y^k}{\partial u^j} + (a_1X^i + a_2Y^i){\Gamma^k}_{ij}\right)\hat{e}_k \\ &= a_1\left(\frac{\partial X^k}{\partial u^j} + X_1^i{\Gamma^k}_{ij}\right)\hat{e}_k + a_2\left(\frac{\partial Y^k}{\partial u^j} + Y^i{\Gamma^k}_{ij}\right)\hat{e}_k \\ &= a_1D_{\hat{e}_j}X + a_2D_{\hat{e}_j}Y \end{aligned} \tag{5.22}$$

Finally,

$$\begin{aligned} D_Z(a_1X + a_2Y) &= D_{Z^j\hat{e}_j}(a_1X + a_2Y) \\ &= Z^jD_{\hat{e}_j}(a_1X + a_2Y) \\ &= a_1Z^jD_{\hat{e}_j}X + a_2Z^jD_{\hat{e}_j}Y \\ &= a_1D_ZX + a_2D_ZY \end{aligned} \tag{5.23}$$

---

3. Prove Eq. (5.6):

$$D_Y(fX) = fD_YX + XD_Yf \tag{5.24}$$

Solution: To prove, Eq. (5.6),

$$\begin{aligned} D_{\hat{e}_j}fX &= \left(\frac{\partial fX^k}{\partial u^j} + fX^i{\Gamma^k}_{ij}\right)\hat{e}_k \\ &= f\left(\frac{\partial X^k}{\partial u^j} + X^i{\Gamma^k}_{ij}\right)\hat{e}_k + X^k\frac{\partial f}{\partial u^j}\hat{e}_k \\ &= fD_{\hat{e}_j}X + X\frac{\partial f}{\partial u^j} \end{aligned} \tag{5.25}$$

Assembling back, we get

$$\begin{aligned} D_YfX &= D_{Y^j\hat{e}_j}fX \\ &= Y^jD_{\hat{e}_j}fX \\ &= Y^j\left(fD_{\hat{e}_j}X + X\frac{\partial f}{\partial u^j}\right) \\ &= \left(fD_YX + XY^j\frac{\partial f}{\partial u^j}\right) \end{aligned} \tag{5.26}$$

---

4. Prove Eq. (5.7):

$$D_Z(X \cdot Y) = (D_Z X) \cdot Y + X \cdot (D_Z Y) \tag{5.27}$$

Solution: To prove Eq. (5.7), we need to use an expression of the Christoffel symbol, which we only state without proof here:

$$\Gamma^k{}_{ij} = \frac{1}{2} g^{ks} \left(\frac{\partial g_{si}}{\partial u^j} + \frac{\partial g_{sj}}{\partial u^i} - \frac{\partial g_{ij}}{\partial u^s} \right) \tag{5.28}$$

Consider the sum of two Christoffel symbols contracted with a metric tensor:

$$\begin{aligned}
\Gamma^k{}_{sj} g_{ki} + \Gamma^k{}_{ij} g_{sk} &= \frac{1}{2} g^{kl} g_{ki} \left(\frac{\partial g_{ls}}{\partial u^j} + \frac{\partial g_{lj}}{\partial u^s} - \frac{\partial g_{sj}}{\partial u^l} \right) \\
&\quad + \frac{1}{2} g^{kl} g_{sk} \left(\frac{\partial g_{li}}{\partial u^j} + \frac{\partial g_{lj}}{\partial u^i} - \frac{\partial g_{ij}}{\partial u^l} \right) \\
&= \frac{1}{2} \delta^l{}_i \left(\frac{\partial g_{ls}}{\partial u^j} + \frac{\partial g_{lj}}{\partial u^s} - \frac{\partial g_{sj}}{\partial u^l} \right) \\
&\quad + \frac{1}{2} \delta^l{}_s \left(\frac{\partial g_{li}}{\partial u^j} + \frac{\partial g_{lj}}{\partial u^i} - \frac{\partial g_{ij}}{\partial u^l} \right) \\
&= \frac{1}{2} \left(\frac{\partial g_{is}}{\partial u^j} + \frac{\partial g_{ij}}{\partial u^s} - \frac{\partial g_{sj}}{\partial u^i} + \frac{\partial g_{si}}{\partial u^j} + \frac{\partial g_{sj}}{\partial u^i} - \frac{\partial g_{ij}}{\partial u^s} \right) \\
&= \frac{\partial g_{is}}{\partial u^j}
\end{aligned} \tag{5.29}$$

Next, begin with the right-hand side of Eq. (5.7) and use $Z = \hat{e}_j$, and then substitute the expression of the covariant derivative and make use of the product rule of the usual calculus:

$$\begin{aligned}
(D_{\hat{e}_j} X) \cdot Y + X \cdot (D_{\hat{e}_j} Y) &= (D_{\hat{e}_j} X)^k g_{ki} Y^i + X^s g_{sk} (D_{\hat{e}_j} Y)^k \\
&= \left(\frac{\partial X^k}{\partial u^j} + X^s \Gamma^k{}_{sj} \right) g_{ki} Y^i \\
&\quad + \left(\frac{\partial Y^k}{\partial u^j} + Y^i \Gamma^k{}_{ij} \right) g_{sk} X^s \\
&= \frac{\partial X^k}{\partial u^j} g_{ki} Y^i + \frac{\partial Y^k}{\partial u^j} g_{sk} X^s \\
&\quad + X^s \Gamma^k{}_{sj} g_{ki} Y^i + Y^i \Gamma^k{}_{ij} g_{sk} X^s \\
&= \frac{\partial X^k}{\partial u^j} g_{ki} Y^i + \frac{\partial Y^k}{\partial u^j} g_{sk} X^s \\
&\quad + X^s Y^i (\Gamma^k{}_{sj} g_{ki} + \Gamma^k{}_{ij} g_{sk})
\end{aligned}$$

Substitute Eq. (5.29) into the last term of the above equation:

$$(D_{\hat{e}_j}X)\cdot Y + X\cdot(D_{\hat{e}_j}Y) = \frac{\partial X^k}{\partial u^j}g_{ki}Y^i + \frac{\partial Y^k}{\partial u^j}g_{sk}X^s + X^sY^i\frac{\partial g_{is}}{\partial u^j}$$

Reindexing and making use of symmetric properties of the metric tensor and using the product rule,

$$\begin{aligned}(D_{\hat{e}_j}X)\cdot Y + X\cdot(D_{\hat{e}_j}Y) &= \frac{\partial X^s}{\partial u^j}g_{si}Y^i + \frac{\partial Y^i}{\partial u^j}g_{si}X^s + X^sY^i\frac{\partial g_{si}}{\partial u^j}\\ &= \frac{\partial X^s g_{si}Y^i}{\partial u^j}\\ &= \frac{\partial}{\partial u^j}(X\cdot Y)\\ &= D_{\hat{e}_j}(X\cdot Y)\end{aligned}$$

Lastly, using the linearity of the covariant derivative with respect to direction, we get

$$D_Z(X\cdot Y) = (D_ZX)\cdot Y + X\cdot(D_ZY)$$

---

5. Prove Eq. (5.8).

$$D_XY - D_YX = [X,Y] = \left(X^i\frac{\partial Y^j}{\partial u^i} - Y^i\frac{\partial X^j}{\partial u^i}\right)\hat{e}_j \tag{5.30}$$

Solution: First, we show this identity with $\hat{e}_j = \delta^k{}_j\hat{e}_k$:

$$\begin{aligned}D_{\hat{e}_i}\hat{e}_j &= D_{\hat{e}_i}\delta^k{}_j\hat{e}_j\\ &= \left(\frac{\partial\delta^k{}_j}{\partial u^i} + \delta^s{}_j\Gamma^k{}_{is}\right)\hat{e}_k\\ &= \Gamma^k{}_{ij}\hat{e}_k\\ &= D_{\hat{e}_j}\hat{e}_i\end{aligned}$$

Using linearity and the product rule of the covariant derivative,

$$\begin{aligned}D_YX - D_XY &= Y^iD_{\hat{e}_i}X^j\hat{e}_j - X^iD_{\hat{e}_i}Y^j\hat{e}_j\\ &= Y^iX^jD_{\hat{e}_i}\hat{e}_j + Y^i\hat{e}_j\frac{\partial X^j}{\partial u^i} - X^iY^jD_{\hat{e}_i}\hat{e}_j + X^i\hat{e}_j\frac{\partial Y^j}{\partial u^i}\\ &= \left(Y^i\frac{\partial X^j}{\partial u^i} - X^i\frac{\partial Y^j}{\partial u^i}\right)\hat{e}_j \qquad (\text{use } D_{\hat{e}_i}\hat{e}_j = D_{\hat{e}_j}\hat{e}_i)\\ &= [Y,X]\end{aligned}$$

5.6 Parallel Transport

Given a vector field on $M \subset \mathbb{R}^n$ and a path $\gamma(t) \in M$, the covariant derivative of the vector field at every point on $\gamma(t)$ in the direction of the path $\dot{\gamma}(t)$ is given in Eq. (5.10). Note that since X is a tangent vector field on M, there is a tangent vector at every point on M. Certainly, there is a tangent vector all along $\gamma(t)$. We can write $X = X(t)$ along $\gamma(t)$. If, among indefinitely many vector fields, we choose a vector field such that

$$D_{\dot{\gamma}} X(t) = 0 \qquad \forall t \in [0, 1] \tag{5.31}$$

then we say that X is being parallel transported along $\gamma(t), t \in [0, 1]$.

Remarks

Parallel transport of a vector on a flat manifold around a closed path will bring the vector back to itself. Parallel transport for a vector on a curved manifold along a closed path will transform the vector into another vector when it returns to the original position. This property is used to compute the curvature of the manifold. Figure 5.1 illustrates the parallel transport of vectors on the 2D Euclidean plane and a 2-sphere.

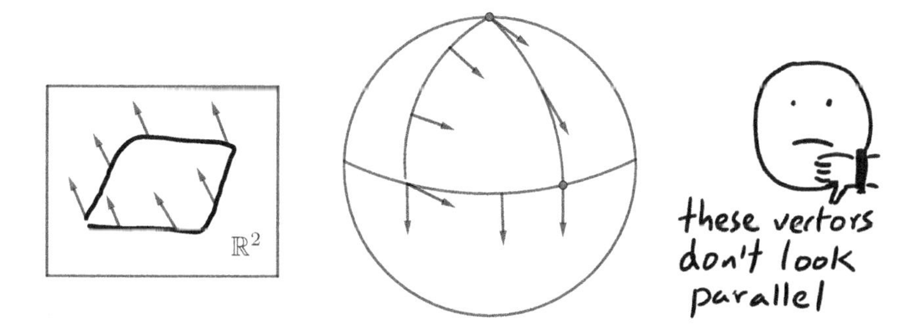

Figure 5.1: Illustration of parallel transport on a 2D plane keeping the direction of the vector invariant. Parallel transport of a vector on a 2-sphere, around the equator, to the north pole, and back again results in the vector turning.

Exercise 5.6.1

1. Given a path on a 2-sphere, parameterize by $u^1 = \theta(t)$ and $u^2 = \phi(t)$:

$$\begin{aligned}\varphi^1(\theta,\phi) &= \cos\phi\sin\theta\\ \varphi^2(\theta,\phi) &= \sin\phi\sin\theta\\ \varphi^3(\theta,\phi) &= \cos\theta\end{aligned}$$

$$\begin{aligned}\theta(t) &= \pi/4\\ \phi(t) &= 2\pi t, \qquad t\in[0,1)\end{aligned}$$

And given a tangent vector,

$$X(0) = \hat{e}_1$$

find the vector field in which $X(0)$ parallel transports along the given path.

Solution: The tangent basis vectors are

$$\begin{aligned}\hat{e}_1 &= \cos\phi\cos\theta e_1 + \sin\phi\cos\theta e_2 - \sin\theta e_3 \qquad (5.32)\\ \hat{e}_2 &= -\sin\phi\sin\theta e_1 + \cos\phi\sin\theta e_2\end{aligned}$$

The path is given by

$$\dot{\gamma}(t) = 2\pi\hat{e}_2(t) \qquad (5.33)$$

For parallel transport, we have

$$D_{\dot{\gamma}}X(t) = 0 \qquad (5.34)$$

$$\begin{aligned}D_{\dot{\gamma}}X(t) &= D_{\hat{e}_1}X\frac{d\theta}{dt} + D_{\hat{e}_2}X\frac{d\phi}{dt} \qquad (5.35)\\ &= 2\pi D_{\hat{e}_2}X\\ &= 0\end{aligned}$$

Using Eq. (5.3),

$$\begin{aligned}\left(\frac{\partial X^k}{\partial\phi} + X^i\Gamma^k{}_{i\phi}\right)\hat{e}_k &= 0 \qquad (5.36)\\ \frac{\partial X^1}{\partial\phi} + X^1\Gamma^1{}_{12} + X^2\Gamma^1{}_{22} &= 0\\ \frac{\partial X^2}{\partial\phi} + X^1\Gamma^2{}_{12} + X^2\Gamma^2{}_{22} &= 0\end{aligned}$$

Substitute in the Christoffel symbols for the 2-sphere using Eq. (4.20):

$$\begin{aligned} \frac{\partial X^1}{\partial \phi} - \cos\theta \sin\theta X^2 &= 0 \\ \frac{\partial X^2}{\partial \phi} + X^1 \frac{\cos\theta}{\sin\theta} &= 0 \end{aligned} \tag{5.37}$$

Using $\theta = \pi/4$ and $\phi = 2\pi t$

$$\begin{aligned} \frac{\partial X^1(t)}{\partial \phi} - \frac{1}{2} X^2(t) &= 0 \\ \frac{\partial X^2(t)}{\partial \phi} + X^1(t) &= 0 \end{aligned} \tag{5.38}$$

Eliminating one equation,

$$\frac{\partial^2 X^1}{\partial \phi^2} = -\frac{1}{2} X^1 \tag{5.39}$$

The general solution is

$$\begin{aligned} X^1 &= A \sin\left(\frac{1}{\sqrt{2}}\phi\right) + B \cos\left(\frac{1}{\sqrt{2}}\phi\right) \\ X^2 &= A\sqrt{2} \cos\left(\frac{1}{\sqrt{2}}\phi\right) - B\sqrt{2} \sin\left(\frac{1}{\sqrt{2}}\phi\right) \end{aligned} \tag{5.40}$$

Solve for A, B using the initial condition:

$$\begin{aligned} X(0) &= (X^1(0), X^2(0)) = (1, 0) \\ \phi(0) &= 2\pi t = 0 \end{aligned} \tag{5.41}$$

We get $A = 0$ and $B = 1$. So,

$$\begin{aligned} X^1(t) &= \cos\left(\sqrt{2}\pi t\right) \\ X^2(t) &= -\sqrt{2} \sin\left(\sqrt{2}\pi t\right) \end{aligned} \tag{5.42}$$

The tangent vector field along the path γ is

$$X(t) = \cos\left(\sqrt{2}\pi t\right) \hat{e}_1 - \sqrt{2} \sin\left(\sqrt{2}\pi t\right) \hat{e}_2 \tag{5.43}$$

Let's look at the direction of the vector when it is parallel transported one round, i.e. $t \to 1$, and compare it with the vector at $t = 0$:

$$\begin{aligned} \lim_{t \to 1} X(t) &= \cos\left(\sqrt{2}\pi\right) \hat{e}_1 - \sqrt{2} \sin\left(\sqrt{2}\pi\right) \hat{e}_2 \\ X(t = 0) &= \hat{e}_1 \end{aligned} \tag{5.44}$$

Note that $\lim_{t \to 1} X(t) \neq X(0)$. Now, solve for solution at arbitrary $\theta(t) = \theta_0$ and for arbitrary θ_0 using Eq. (5.37):

$$\frac{\partial^2 X^1}{\partial \phi^2} = -(\cos^2 \theta_0) X^1 \tag{5.45}$$

$$X^1 = A \sin(2\pi t \cos \theta_0) + B \cos(2\pi t \cos \theta_0) \tag{5.46}$$

Using the initial condition $A = 0, B = 1$,

$$X(t) = \cos(2\pi t \cos \theta_0) \hat{e}_1 - \frac{\sin(2\pi t \cos \theta_0)}{\sin \theta_0} \hat{e}_2 \tag{5.47}$$

For parallel transport, the vector "does not turn" in the manifold's local coordinates in $\mathbb{R}^3$; however, the vector turns. Let's see how much this vector varies in $\mathbb{R}^3$ when it gets transported one whole round and reaches its original point:

$$\begin{aligned} X(0) \cdot X(1) &= \cos(2\pi \cos \theta_0) \hat{e}_1 \cdot \hat{e}_1 - \frac{\sin(2\pi \cos \theta_0)}{\sin \theta_0} \hat{e}_2 \cdot \hat{e}_1 \\ &= \cos(2\pi \cos \theta_0) \end{aligned} \tag{5.48}$$

We see that when $\theta_0 = \pi/2$, $X(0) \cdot X(1) = 1$.

5.6.1 Properties of parallel transport

Lemma 5.6.1 *These properties of parallel transport hold:*

1. *Given two vector fields $X(t)$ and $Y(t)$, in which they are parallel transported along a path $\gamma(t)$ in M, then their inner product is constant:*

$$X(0) \cdot Y(0) = X(t) \cdot Y(t) \quad \forall t \tag{5.49}$$

2. *Given a path $\gamma(t)$, if*

$$D_{\dot{\gamma}(t)} \dot{\gamma}(t) = 0 \quad \forall t \tag{5.50}$$

then the path $\gamma(t)$ traces out a geodesic.

Exercise 5.6.2

1. Show that if X and Y undergo parallel transport along the same path γ, then

$$\frac{dX(t) \cdot Y(t)}{dt} = 0$$

This is also a statement of Eq. (5.49).

Solution: To prove Eq. (5.49), we need to show that

$$\frac{dX(t) \cdot Y(t)}{dt} = 0$$

The equations for parallel transport for X and Y are

$$\begin{aligned} D_{\dot{\gamma}}X &= (D_{\hat{e}_j}X)\frac{du^j}{dt} = 0 \\ D_{\dot{\gamma}}Y &= (D_{\hat{e}_j}Y)\frac{du^j}{dt} = 0 \end{aligned}$$

Writing out in components, we get

$$\begin{aligned} \frac{\partial X^k}{\partial u^j}\frac{du^j}{dt} + X^i\Gamma^k{}_{ij}\frac{du^j}{dt} &= 0 \\ \frac{\partial Y^s}{\partial u^j}\frac{du^j}{dt} + Y^i\Gamma^s{}_{ij}\frac{du^j}{dt} &= 0 \end{aligned}$$

Multiply the top equation by $g_{sk}Y^s$ and the bottom equation by $g_{sk}X^k$ and write the first term of the above equations as total derivatives:

$$\begin{aligned} g_{sk}Y^s\frac{dX^k}{dt} + g_{sk}Y^sX^i\Gamma^k{}_{ij}\frac{du^j}{dt} &= 0 \\ g_{sk}X^k\frac{dY^s}{dt} + g_{sk}X^kY^i\Gamma^s{}_{ij}\frac{du^j}{dt} &= 0 \end{aligned}$$

Reindex the second term of the second equation by $s \to k, k \to i, i \to s$:

$$g_{sk}X^kY^i\Gamma^s{}_{ij}\frac{du^j}{dt} \to g_{ki}X^iY^s\Gamma^k{}_{sj}\frac{du^j}{dt}$$

After reindexing, add the two equations:

$$\begin{aligned} g_{sk}Y^s\frac{dX^k}{dt} + g_{sk}Y^sX^i\Gamma^k{}_{ij}\frac{du^j}{dt} &= 0 \\ g_{sk}X^k\frac{dw^s}{dt} + g_{ki}X^iY^s\Gamma^k{}_{sj}\frac{du^j}{dt} &= 0 \end{aligned}$$

Adding and using the product rule gives

$$\begin{aligned}
\frac{dX\cdot Y}{dt} - Y^sX^k\frac{dg_{sk}}{dt} + g_{sk}Y^sX^i\Gamma^k{}_{ij}\frac{du^j}{dt} + g_{ki}X^iY^s\Gamma^k{}_{sj}\frac{du^j}{dt} &= 0\\
\frac{dX\cdot Y}{dt} - Y^sX^k\frac{dg_{sk}}{dt} + Y^sX^i\left(g_{sk}\Gamma^k{}_{ij} + g_{ki}\Gamma^k{}_{sj}\right)\frac{du^j}{dt} &= 0
\end{aligned}$$

The term in parenthesis is given by Eq. (5.29) as a partial derivative of the metric tensor:

$$\frac{dX\cdot Y}{dt} - Y^sX^k\frac{dg_{sk}}{dt} + Y^sX^i\frac{\partial g_{is}}{\partial u^j}\frac{du^j}{dt} = 0$$

Reindex $i \to k$ and convert the partial derivative of the metric tensor to a total derivative:

$$\begin{aligned}
\frac{dX\cdot Y}{dt} - Y^sX^k\frac{dg_{sk}}{dt} + Y^sX^k\frac{dg_{ks}}{dt} &= 0\\
\frac{dX\cdot Y}{dt} &= 0
\end{aligned}$$

---

2. Prove that Eq. (5.50) implies γ is a geodesic.

Solution: To prove Eq. (5.50), begin from the usual derivative:

$$\begin{aligned}
\frac{d\dot{\gamma}}{dt} &= \frac{d\dot{\gamma}^j\hat{e}_j}{dt} \qquad (5.51)\\
&= \frac{d\dot{\gamma}^j}{dt}\hat{e}_j + \dot{\gamma}^j\frac{d\hat{e}_j}{dt}\\
&= \ddot{\gamma}^j\hat{e}_j + \dot{\gamma}^j\frac{\partial\hat{e}_j}{\partial u^i}\frac{du^i}{dt}
\end{aligned}$$

Differentiate in the direction of γ; hence,

$$\frac{du^i}{dt} = \dot{\gamma}^i \qquad (5.52)$$

$$\frac{d\dot{\gamma}}{dt} = \left(\ddot{\gamma}^k + \dot{\gamma}^j\dot{\gamma}^i\Gamma^k{}_{ij}\right)\hat{e}_k + \text{normal components} \qquad (5.53)$$

The covariant derivative takes the tangential direction of the usual derivative, and given that the covariant derivative is zero,

$$D_{\dot{\gamma}}\dot{\gamma} = \left(\ddot{\gamma}^k + \dot{\gamma}^j\dot{\gamma}^i\Gamma^k{}_{ij}\right)\hat{e}_k = 0 \qquad (5.54)$$

Recalling the geodesic equation in Eq. (4.13), we see that the covariant derivative along $\dot{\gamma}$ being zero leads to the geodesic equation.

5.6.2 Exponential map

Consider a point $x \in M$ and a tangent vector $X \in T_xM$. One could find a unique geodesic by solving the geodesic equation with the required initial conditions:

$$\begin{aligned} \ddot{\gamma}^k + \dot{\gamma}^j\dot{\gamma}^i\Gamma^k{}_{ij} &= 0 \\ \gamma(0) &= x \\ \dot{\gamma}(0) &= X \end{aligned} \tag{5.55}$$

Let $\gamma^*(t)$ be the solution to Eq. (5.58); this solution will be unique. An exponential map is defined as

$$\exp(X) = \gamma^*(1) \tag{5.56}$$

5.6.3 Scaling properties of geodesics

Let $\gamma(t)$ be a geodesic such that

$$D_{\dot{\gamma}}\dot{\gamma} = 0$$

Then,

$$\dot{\gamma}(\alpha t) = \alpha\dot{\gamma}(t)$$
$$\alpha^2 D_{\dot{\gamma}}\dot{\gamma} = D_{\alpha\dot{\gamma}}\alpha\dot{\gamma} = 0$$

If we reparameterize $\tau = \alpha t$, then $\gamma(\tau)$ is a geodesic with the initial conditions $\dot{\gamma}(\tau) = \alpha\dot{\gamma}(t)$.

Exercise 5.6.3

1. Given any path in M, it can be parameterized by $\gamma(t)$ as well as $\gamma(2t)$ (or $\gamma(\alpha t)$ for any α). All parameterizations $\gamma(\alpha t), t \in \mathbb{R}$ give the same trace, with the only difference being the speed of the trace. Show that

$$\begin{aligned} \dot{\gamma}(\alpha t) &= \alpha\dot{\gamma}(t) \\ \ddot{\gamma}(\alpha t) &= \alpha^2\ddot{\gamma}(t) \end{aligned} \tag{5.57}$$

If the above equations hold, then substitute them into the geodesic expression:

$$\begin{aligned} \ddot{\gamma}^k(\alpha t) + \dot{\gamma}^j(\alpha t)\dot{\gamma}^i(\alpha t)\Gamma^k{}_{ij} &= \alpha^2\ddot{\gamma}^k(t) + \alpha^2\dot{\gamma}^j(t)\dot{\gamma}^i(t)\Gamma^k{}_{ij} \\ \alpha^2\left(\ddot{\gamma}^k(t) + \dot{\gamma}^j(t)\dot{\gamma}^i(t)\Gamma^k{}_{ij}\right) &= 0 \\ \gamma(0) &= x \\ \dot{\gamma}(0) &= X \end{aligned} \tag{5.58}$$

It seems that $\gamma(\alpha t)$ also satisfy the geodesic equation. However, by the uniqueness theorem of second-order differential equations, we can only have one solution. What is the contradiction here? (*Hint*: Work out a counterexample.)

Solution:

$$\dot{\gamma}(\alpha t) = \frac{d\gamma(\alpha t)}{dt} = \alpha\frac{d\gamma(\alpha t)}{d\alpha t} = \alpha\frac{d\gamma(\tau)}{d\tau} = \alpha\dot{\gamma}(t)$$

Applying the same procedure again to $\dot{\gamma}(t)$:

$$\ddot{\gamma}(\alpha t) = \frac{d^2\gamma(\alpha t)}{dt^2} = \alpha^2\frac{d\gamma(\alpha t)}{d(\alpha t)^2} = \alpha^2\frac{d\dot{\gamma}(\tau)}{d\tau} = \alpha\ddot{\gamma}(t)$$

Solutions to second-order differential equations also depend on the initial conditions. When the initial conditions are different, the solutions are different. One could solve the geodesic equation using $\dot{\gamma}(0) = X$ or $\dot{\gamma}(0) = \alpha X$. The resulting geodesic solutions will be $\gamma(t)$ and $\gamma(\alpha t)$.

5.7 Summary

The derivative of a vector field is given by

$$\begin{aligned}\frac{\partial X}{\partial u^j} &= \frac{\partial X^i}{\partial u^j}\hat{e}_i + X^i\frac{\partial \hat{e}_i}{\partial u^j} \\ &= \left(\frac{\partial X^k}{\partial u^j} + X^i\Gamma^k{}_{ij}\right)\hat{e}_k + X^i L^k{}_{ij}\hat{n}_k\end{aligned} \tag{5.59}$$

By dropping the second fundamental form terms, we get the covariant derivative.

Definition 5.7.1 The covariant derivative of a vector Y in the direction of X is given by

$$D_X Y = X^j\left(\frac{\partial Y^k}{\partial u^j} + Y^i\Gamma^k{}_{ij}\right)\hat{e}_k \in T_x M \tag{5.60}$$

Definition 5.7.2 The covariant derivative on a path $\gamma(t)$ is given by

$$D_{\dot{\gamma}} X = \left(D_{\hat{e}_j} X\right)\frac{du^j}{dt} = \left(D_{\hat{e}_j} X\right)\dot{\gamma}^j \tag{5.61}$$

Lemma 5.3.1 *Let $D_Y X$ be the covariant derivative of X in the direction $Y \in T_x M$. Let $Z \in T_x M$ and $X \in T_x M$. Then, the following identities hold:*

1. *Linear in the direction of derivative: $f, g : M \mapsto \mathbb{R}$.*

$$D_{fY_1+gZ_2}X = fD_Y X + gD_Z X \tag{5.4}$$

2. *Linear in input: $a_1, a_2 \in \mathbb{R}$.*

$$D_Z(a_1X + a_2Y) = a_1D_Z X + a_2D_Z Y \tag{5.5}$$

3. *Product rule, $f : M \mapsto \mathbb{R}$:*

$$D_Y(fX) = fD_Y X + XD_Y f \tag{5.6}$$

4. *Metric compatibility:*

$$D_Z(X \cdot Y) = (D_Z X) \cdot Y + X \cdot (D_Z Y) \tag{5.7}$$

5. *Torsion-free property:*

$$D_X Y - D_Y X = [X, Y] = \left(X^i \frac{\partial Y^j}{\partial u^i} - Y^i \frac{\partial X^j}{\partial u^i} \right) \hat{e}_j \tag{5.8}$$

Definition 5.7.3 Given a path $\gamma(t)$, a vector field $X(t)$ along $\gamma(t)$ is being parallel transported if

$$D_{\dot{\gamma}} X(t) = 0 \qquad \forall t \in [0, 1] \tag{5.62}$$

Lemma 5.6.1 *These properties of parallel transport hold:*

1. *Given two vector fields $X(t)$ and $Y(t)$, in which they are parallel transported along a path $\gamma(t)$ in M, then their inner product is constant:*

$$X(0) \cdot Y(0) = X(t) \cdot Y(t) \qquad \forall t \tag{5.49}$$

2. *Given a path $\gamma(t)$, if*

$$D_{\dot{\gamma}(t)} \dot{\gamma}(t) = 0 \qquad \forall t \tag{5.50}$$

then the path $\gamma(t)$ traces out a geodesic.

Definition 5.7.4 Given a geodesic $\gamma(t)$ with $\gamma(0) = x$, $\dot{\gamma}(0) = X$, the exponential map is a map $\exp : T_x M \mapsto M$ given by

$$\exp(X) = \gamma(1) \tag{5.63}$$

Lemma 5.7.1 *If $\gamma(t)$ be a geodesic such that*

$$D_{\dot{\gamma}} \dot{\gamma} = 0 \tag{5.64}$$

and reparameterizing $\tau = \alpha t$, then $\gamma(\tau)$ is a geodesic with $\dot{\gamma}(\tau) = \alpha \dot{\gamma}(t)$.

Chapter 6

Intrinsic Geometry

The previous chapters assumed a manifold embedded in $\mathbb{R}^n$. If we wish to use differential geometry to model our physical world, we should challenge the assertion that our world is Euclidean in nature. The theory of general relativity tells us that space-time is curved. If this is so, everything in our universe is curved, and hence we cannot measure things in $\mathbb{R}^n$. However, we can assume that, at a "local" level, our space is approximately Euclidean, e.g. the space in a classroom.

This locality concept is interesting. Before we knew that the Earth is a globe, in our perspective, the land is flat globally except for hills and valleys. Suppose one is able to measure the longitude and latitude by observing the stars. One could construct a coordinate representation of longitudes and latitudes. Then, one will find that the distance between the longitude is shorter nearer to the north and south poles. Indeed, in this space, the metric tensor is not the same as the metric tensor for the Euclidean $\mathbb{R}^2$. Without knowledge of global geometry, i.e. the globe embedded in $\mathbb{R}^3$, how can one compute geodesics? Indeed, geodesics can be computed using the metric tensor alone. There is no need for knowledge of the embedded space. In this chapter, we discuss how to compute various geometrical quantities without the use of embedded space.

6.1 A Summary of Extrinsic Geometry

We first give a summary of formalism in the previous chapters. We have considered a manifold embedded in $\mathbb{R}^n$; when this happens, differentiation in the usual calculus sense becomes ill-defined because we have no access to all points in $\mathbb{R}^n$ that are outside of the manifold. In particular, the manifold is usually a hypersurface. To circumvent this issue, we define a chart that maps from some lower-dimensional space U to points in the manifold. Points on the manifold have coordinates in $\mathbb{R}^n$, so the chart maps $U \mapsto \mathbb{R}^n$:

$$\varphi : U \mapsto \mathbb{R}^n$$

We have used $U \subset \mathbb{R}^d$, where d is the dimension of the manifold. In this way, any function that maps from the manifold to some other space can be differentiated with respect to U:

$$f : M \mapsto N$$

Then, differentiation uses the coordinate representation u:

$$\frac{\partial f(\varphi(u))}{\partial u^i}$$

Tangent vectors are defined as

$$\hat{e}_i(u) = \frac{\partial \varphi(u)}{\partial u^i}$$

Note that the tangent vectors are functions of u. The metric tensor is one that is induced from the Euclidean space $\mathbb{R}^n$ by taking the inner product of tangent vectors:

$$g_{ij}(u) = \hat{e}_i(u) \cdot \hat{e}_j(u)$$

The dot product in this equation is the dot product in the embedded Euclidean space. The metric tensor is also a function of u.

6.2 From Extrinsic to Intrinsic Geometry

This section discusses how to transition from extrinsic to intrinsic geometry. In extrinsic geometry, we have the manifold M embedded in $\mathbb{R}^n$. There exists a chart $\varphi : U \mapsto M$. From this mapping, we define tangent basis vectors and therefore the metric tensor. The key points of this transition are listed as follows:

1. We no longer use the mapping $\varphi(u)$.

2. Instead of defining two spaces U and M related by φ, we only use one space in intrinsic geometry, and we only use U and consider it as the manifold.

3. The tangent vectors defined in extrinsic geometry can no longer be defined because there are no position vectors $\varphi(u)$ to differentiate with. An alternative way of expressing tangent vectors is by partial derivatives in U:

$$X \equiv X^i \frac{\partial}{\partial u^i} = X^i \partial_{u^i}$$

4. The metric tensor cannot be derived using the method of induced metric tensor in extrinsic geometry. In intrinsic geometry, we assume that the metric tensor is given or derived from other means. Furthermore, we assume the inverse of the metric tensor exists and

$$g^{ij} g_{jk} = \delta^i{}_k$$

5. The Christoffel symbols are derived from the metric tensor as

$$\Gamma^k{}_{ij} = \frac{1}{2} g^{ks} \left(\frac{\partial g_{si}}{\partial u^j} + \frac{\partial g_{sj}}{\partial u^i} - \frac{\partial g_{ij}}{\partial u^s} \right)$$

6.2.1 Tangent vectors for intrinsic geometry

We no longer use the manifold parameterization mapping $\varphi(u)$ in intrinsic geometry; therefore, the tangent vectors which we defined in extrinsic geometry as derivatives of φ, i.e. $\hat{e}_j = \partial\varphi(u)/\partial u^j$, no longer exist. A workaround for the formal definition of tangent vectors is to associate the differential operators with tangent basis vectors:

$$\hat{e}_i \equiv \frac{\partial}{\partial u^i} = \partial_{u^i} \tag{6.1}$$

However, from a physics perspective, a vector has a meaning akin to velocity, force, etc. While differential operators have their meaning in the sense of a gradient. In this perspective, differential operators are not, in an intuitive sense, vectors. We still need to reconcile with what is generally accepted in the literature. First, let's state some facts:

1. If the differential operator is being "fed" with the position vector on M in extrinsic coordinates, it becomes the basis for tangent vectors:

$$\frac{\partial}{\partial u^i} \varphi = \hat{e}_i$$

2. Differentials transform the same way as basis vectors; they transform covariantly:

$$\frac{\partial}{\partial \hat{u}^j} = J^k{}_j \frac{\partial}{\partial u^k}$$

3. Differentials can be used on a function that maps a point on U to $\mathbb{R}$[1]:

$$f : U \mapsto \mathbb{R}$$

Then, given a vector in $T_u U$, $X = X^i \partial_{u^i}$, its action on a function is given by the usual calculus differential:

$$X(f) = X^i \frac{\partial}{\partial u^i} f(u)$$

[1]Differentiation is valid only with coordinate representation.

Formalism of the new tangent vector

In explicit geometry, we denote M as the manifold, while in intrinsic geometry, we use a different symbol: we denote U as the manifold and T_uU as the tangent space.

Definition 6.2.1 (tangent vectors) Given a manifold U, with $u \in U$, define the basis tangent vector as

$$\hat{e}_i = \frac{\partial}{\partial u^i} = \partial_{u^i}$$

Tangent vectors are given by $X = X^i \partial_{u^i}$, which is a linear mapping

$$X : C^\infty(U) \mapsto \mathbb{R}$$

C^∞ is the set of differentiable functions that maps U to $\mathbb{R}$.

We can define tangent vectors in the context of a path on a manifold. Recall that for explicit geometry, we use the chain rule to get the components of tangent vectors:

$$\frac{d\gamma(t)}{dt} = \frac{d\varphi(u(t))}{dt} = \frac{du^i(t)}{dt}\frac{\partial\varphi(u)}{\partial u^i} = X^i \hat{e}_i$$

For intrinsic geometry, we do not have the function $\varphi(u(t))$ and $\gamma(t) = u(t)$; when we differentiate γ, we do not get the basis tangent vectors. The mathematics does not work out! We can only conclude that, for intrinsic geometry, we cannot consider tangent vectors from a geometrical point of view. This conclusion is consistent with the discussion earlier in this section that we cannot consider differential operators with a physical or geometrical interpretation. We can only define tangent vectors similarly to Eq. (6.1):

$$\frac{d\gamma(t)}{dt} \equiv X^i \hat{e}_i = X^i \partial_{u^i}$$

Definition 6.2.2 Tangent vectors can be defined as the derivative of a path on the manifold:

$$\frac{d\gamma(t)}{dt} \equiv \frac{du^i}{dt}\frac{\partial}{\partial u^i} = X^i \partial_{u^i} \tag{6.2}$$

Remarks

We discuss how the simple geometrical interpretation of tangent vectors does not work out. In intrinsic geometry, a point in the manifold is given by u, and a path is the parametrization of points $u(t)$. Then, $\gamma(t) = u(t)$. A differential with respect to time is given by

$$\frac{d\gamma(t)}{dt} = \frac{du(t)}{dt}$$

The above equation is inconsistent with Eq. (6.2). It has a geometric interpretation in U as it specifies the direction of the path in U. Indeed, the components in the U coordinate system are

$$X^i = \frac{du^i(t)}{dt}$$

However, we are missing the basis vectors. If we consider the basis vectors as ∂_{u^i}, we can concatenate them with the components:

$$X = X^i \partial_{u^i}$$

Tangent vector fields

We define a tangent vector at every point in the manifold U. With the definitions of tangent vector, given a scalar function $f : U \mapsto \mathbb{R}$,

$$X(u_0)(f) - X^i(u_0) \frac{\partial f(u)}{\partial u^i}\bigg|_{u_0}$$

The tangent vector above is evaluated at the point u_0. We can drop the subscript in cases when the point of evaluation is arbitrary and does not cause any confusion. For the case of tangent vectors derived from a path at an arbitrary point (and therefore dropping the subscript),

$$X(f) = \frac{df(u)}{dt} = \frac{du^i}{dt}\frac{\partial f(u)}{\partial u^i} = \dot{\gamma}\frac{\partial f(u)}{\partial u^i}$$

6.2.2 Abstract definitions of tangent vectors

We introduce one more level of abstraction for tangent vectors by defining them as objects with certain properties:

1. With $\alpha, \beta \in \mathbb{R}$, $f, g \in C^\infty(U)$, $X \in T_u U$,

$$X(\alpha f + \beta g) = \alpha X(f) + \beta X(g)$$

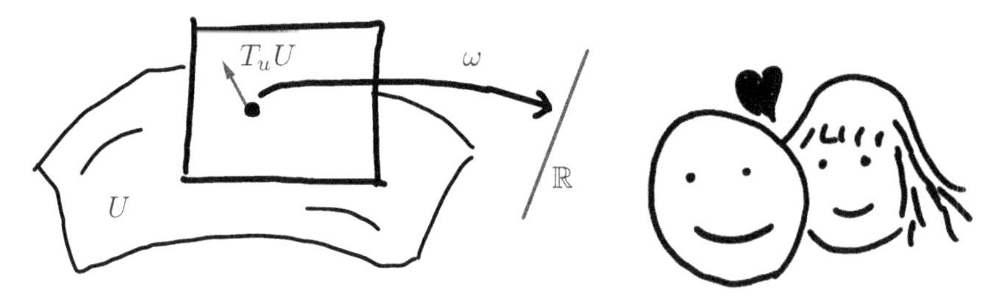

Figure 6.1: Covector as a mapping $T_uU \mapsto \mathbb{R}$.

2. With $\alpha, \beta \in \mathbb{R}$, $f \in C^\infty(U)$, $X, Y \in T_uU$,

$$(\alpha X + \beta Y)(f) = \alpha X(f) + \beta Y(f)$$

3. With $\alpha, \beta \in \mathbb{R}$, $f, g \in C^\infty(U)$, $X \in T_uU$,

$$X(fg) = f(u)X(g) + X(f)g(u)$$

6.3 Covectors

With tangent vectors defined as operators on scalar functions, we can define covectors (differential 1-form) that take in a tangent vector and return a scalar. As we will see later, covectors are duals of tangent vectors. We can bring covectors to another level of abstraction by just defining how they operate on vectors, but we will do this later. First, we define covectors using differential operators for tangent vectors. Consider a function:

$$f : U \mapsto \mathbb{R}$$

Define the differential of f as some object that takes in a tangent vector and outputs a scalar:

$$\begin{aligned} df &: T_uU \mapsto \mathbb{R} \\ df(X) &= X(f) = X^i \partial_{u^i} f \in \mathbb{R} \end{aligned} \tag{6.3}$$

f takes in u and gives $\mathbb{R}$, $f = f(u^1, \ldots, u^d)$. An example of f can be $f(u) = \sum_i (u^i)^2$.

If we let $f = u^k$ and $X = \partial_{u^i}$, then

$$df(X) = du^k(\partial_{u^i}) = \frac{\partial u^k}{\partial u^i} = \delta^k{}_i \tag{6.4}$$

We see that the covector du^k is orthogonal to the tangent vector ∂_{u^j}. Figure 6.1 illustrates the mapping of tangent vector by a covector onto $\mathbb{R}$. We give a formal definition of a covector here.

Definition 6.3.1 (Covectors) Given a scalar function $f : U \mapsto \mathbb{R}$, define its differential df as a map $df : T_u U \mapsto \mathbb{R}$ as

$$df(X) = X(f) \qquad \forall X \in T_u U$$

6.3.1 Covector basis

We may ask the question of what are all possible covectors and what is the dimension of this space. From Eq. (6.4), for every basis tangent vectors ∂_{u^i}, there is a covector du^i. ∂_{u^i} span $T_u U$, and any tangent vector can be written as $X = X^i \partial_{u^i}$; the same is true for covectors:

$$\omega = \omega_i du^i$$

for some ω_i. Indeed, du^i forms a dual basis of ∂_{u^i}. Hence, we call the space spanned by covectors the dual space of the tangent space:

$$\omega \in T_u^* U$$

ω_i can be uniquely defined if $\omega = df$:

$$\begin{aligned} df\left(\partial_{u^j}\right) &= \partial_{u^j} f \\ \omega_i du^i\left(\partial_{u^j}\right) &= \omega_j = \partial_{u^j} f \end{aligned}$$

Therefore, a differential of a function is given by

$$df = (\partial_{u^i} f) du^i$$

This expression is similar to the chain rule in the usual calculus expression that we are familiar with:

$$df = \frac{\partial f}{\partial u^1} du^1 + \frac{\partial f}{\partial u^2} du^2 + \cdots + \frac{\partial f}{\partial u^d} du^d$$

$$df(X) = \partial_{u^i} f du^i(X) = X^i \partial_{u^i} f$$

We see that covectors evaluate to be the direction derivative.

6.3.2 Abstract definition of covectors

Once we define covectors abstractly, it no longer matters if tangent vectors are defined as a partial differential ∂_{u^i} or as a geometric vector as in extrinsic geometry $\partial_{u^i} \varphi$. We define abstract objects e_i as basis vectors instead of in explicit form as derivatives.

Let T_uU be a vector space with basis e_i, $i = 1, \dots, d$, and its dual space T_uU^* has basis $\tilde{e}^i$, $i = 1, \dots, d$ such that

$$\tilde{e}^i(e_j) = \delta^i{}_j \tag{6.5}$$

$\tilde{e}^i$ are basis covectors, and we define them to be linear. For $\alpha, \beta \in \mathbb{R}$,

$$\tilde{e}^i(\alpha e_j + \beta e_k) = \alpha\tilde{e}^i(e_j) + \beta\tilde{e}^i(e_k)$$

$$(\alpha\tilde{e}^i + \beta\tilde{e}^k)(e_j) = \alpha\tilde{e}^i(e_j) + \beta\tilde{e}^k(e_j)$$

Recall that tangent vectors transform covariantly as

$$\hat{e}_i = J^j{}_i e_j$$

It can be shown that covectors transform contravariantly. Using the assumption of coordinate independence of covectors,

$$\omega_i\tilde{e}^i = \hat{\omega}_i\hat{\tilde{e}}^i$$

Applying to e_j on both sides,

$$\begin{aligned} \omega_i\tilde{e}^i(e_j) &= \hat{\omega}_i\hat{\tilde{e}}^i(e_j) \\ \omega_j &= \hat{\omega}_i\hat{\tilde{e}}^i(J^{-1k}{}_j\hat{e}_k) \end{aligned}$$

Using the linearity property,

$$\begin{aligned} \omega_j &= J^{-1k}{}_j\hat{\omega}_i\hat{\tilde{e}}^i(\hat{e}_k) \\ \omega_j &= J^{-1i}{}_j\hat{\omega}_i \end{aligned}$$

Therefore,

$$\begin{aligned} \hat{\omega}_i &= J^j{}_i\omega_j \\ \hat{\tilde{e}}^i &= J^{-1i}{}_j\tilde{e}^j \end{aligned}$$

6.3.3 Covectors in extrinsic geometry

In extrinsic geometry, we have the manifold $M \subset \mathbb{R}^n$ and the chart $\varphi : U \mapsto M$. Let φ^k be the kth component of the image of φ. The vector space in $\mathbb{R}^n$ is spanned by $e_i, i = 1, \dots, n$ and the dual space by $\tilde{e}_i$. Note that $\varphi^k : U \mapsto \mathbb{R}$ is a scalar function. Define the differential of φ^k as

$$d\varphi^k(e_j) = \delta^k{}_j \tag{6.6}$$

with the usual linearity properties of a covector. With this definition, we can evaluate the action of $d\varphi^k$ on a tangent vector in extrinsic geometry representation. Let's work out the covector's action on tangent vectors for the 2-sphere:

$$\begin{aligned}\varphi^1 &= \sin\theta\cos\phi \\ \varphi^2 &= \sin\theta\sin\phi \\ \varphi^3 &= \cos\theta\end{aligned}$$

Assigning $u^1 = \theta, u^2 = \phi$, the tangent vectors are

$$\begin{aligned}\hat{e}_1 &= \cos\theta\cos\phi e_1 + \cos\theta\sin\phi e_2 - \sin\theta e_3 \\ \hat{e}_2 &= -\sin\theta\sin\phi e_1 + \sin\theta\cos\phi e_2\end{aligned}$$

Using Eq. (6.6), the covector just picks up its corresponding components of the tangent vectors:

$$\begin{aligned}d\varphi^1(\hat{e}_1) &= \cos\theta\cos\phi \\ d\varphi^2(\hat{e}_1) &= \cos\theta\sin\phi \\ d\varphi^3(\hat{e}_1) &= -\sin\theta \\ d\varphi^1(\hat{e}_2) &= -\sin\theta\sin\phi \\ d\varphi^2(\hat{e}_2) &= \sin\theta\cos\phi \\ d\varphi^3(\hat{e}_2) &= 0\end{aligned}$$

We derive here a general expression for applying covectors on tangent vectors:

$$\begin{aligned}\hat{e}_i &= (\partial_{u^i}\varphi^j)e_j \\ d\varphi^k(\hat{e}_i) &= d\varphi^k((\partial_{u^i}\varphi^j)e_j) \\ &= (\partial_{u^i}\varphi^j)d\varphi^k(e_j) \\ &= (\partial_{u^i}\varphi^j)\delta^k{}_j \\ &= \partial_{u^i}\varphi^k\end{aligned}$$

So, a covector acting on a general tangent vector $X = X^i e_i^u$ is

$$d\varphi^k(X) = X^i d\varphi^k(\hat{e}_i) = X^i\partial_{u^i}\varphi^k$$

Note that $\partial_{u^i}\varphi^k$ is the kth component of the ith tangent basis vector. Covectors can be written as

$$\omega = \omega_i d\varphi^i \qquad \text{instead of} \qquad \omega_i du^i$$

Table 6.1: Comparison of tangent vectors and differential forms in extrinsic and intrinsic geometry perspectives.

	Extrinsic geometry	Intrinsic geometry
Tangent vectors	$\varphi : U \mapsto M$	no parametric mapping
	$\hat{e}_j = (\partial_{u^j} \varphi^k) e_k$	$\hat{e}_j = \partial_{u^j}$
Covectors	no explicit form	$f : U \mapsto \mathbb{R}$, $df(X) = X(f)$
	$d\varphi^k(e_j) = \delta^k{}_j$, $d\varphi^k(\hat{e}_j) = \partial_{u^j} \varphi^k$	$du^i(e^u_j) = \partial_{u^j} u^i = \delta^i{}_j$

To understand how to make the transition between intrinsic and extrinsic geometry views, we need to revise how to make the transition between these two views for tangent vectors. In Table 6.1, we compare and contrast tangent vectors and covectors in extrinsic and intrinsic geometry.

Exercise 6.3.1

1. Given $f(u^1, u^2) = \cos(u^2)\sin(u^1) + \frac{1}{2}\sin^2(u^2)$, express df in terms of u^i and du^i.

Solution:

$$
\begin{aligned}
df &= \frac{\partial f}{\partial u^1} du^1 + \frac{\partial f}{\partial u^2} du^2 \\
\frac{\partial f}{\partial u^1} &= \cos u^1 \cos u^2 \\
\frac{\partial f}{\partial u^2} &= -\sin u^2 \sin u^1 + \sin u^2 \cos u^2 \\
&= \sin u^2 (\cos u^2 - \sin u^1)
\end{aligned}
$$

$$df = \cos u^1 \cos u^2 du^1 + \sin u^2 (\cos u^2 - \sin u^1) du^2$$

---

2. Given $f(u^1, u^2) = \exp(u^1) + u^2$ and tangent vector field $X = (u^2)^2 \partial/\partial u^1 + u^1 \partial/\partial u^2$, evaluate $df(X)$.

Solution:

$$
\begin{aligned}
df(X) = X(f) &= (u^2)^2 \frac{\partial}{\partial u^1}(\exp(u^1) + u^2) + u^1 \frac{\partial}{\partial u^2}(\exp(u^1) + u^2) \\
&= (u^2)^2 \exp(u^1) + u^1
\end{aligned}
$$

---

3. Prove the properties of covectors: linearity, $df(X + Y) = df(X) + df(Y)$, $d(f + g) = df + dg$, and the product rule, $d(fg) = f dg + g df$.

Solution:

$$\begin{aligned} df(X+Y) &= (X+Y)(f) = X(f) + Y(f) \\ d(f+g)(X) &= X(f+g) = X(f) + X(g) = df(X) + dg(X) \end{aligned}$$

X is arbitrary; therefore, $d(f+g) = d(f) + d(g)$.

$$d(fg)(X) = X(fg) = fX(g) + gX(f) = fdg(X) + gdf(X)$$

X is arbitrary; therefore, $dfg = fdg + gdf$.

6.3.4 Covariant derivative of covectors

Given a covector ω operating on a tangent vector X and using the Leibniz rule[2] for the covariant derivative,

$$D_{\partial_{u^i}}\omega(X) = \omega(D_{\partial_{u^i}}X) + (D_{\partial_{u^i}}\omega)(X) \tag{6.7}$$

The left-hand side of Eq. (6.7) is a function, and the covariant becomes a directional derivative:

$$\frac{\partial \omega_j X^j}{\partial u^i} = \omega_j \frac{\partial X^j}{\partial u^i} + X^j \frac{\partial \omega_j}{\partial u^i} \tag{6.8}$$

Write the covector as $\omega = \omega_j du^j$ and the vector as $X = X^j \partial_{u^j}$. Then, the first term on the right-hand side of Eq. (6.7) is

$$\begin{aligned} \omega_j du^j \left(\frac{\partial X^k}{\partial u^i} + X^s {\Gamma^k}_{is} \right) \partial_{u^k} &= \left(\omega_j \frac{\partial X^k}{\partial u^i} + \omega_j X^s {\Gamma^k}_{is} \right) du^j(\partial_{u^k}) \\ &= \left(\omega_j \frac{\partial X^j}{\partial u^i} + \omega_j X^s {\Gamma^j}_{is} \right) \end{aligned} \tag{6.9}$$

The covariant derivative of a tangent vector is a tangent vector. By the Leibniz rule, the covariant derivative of a covector is a covector. Let $\alpha = \alpha_j du^j = D_{\partial_{u^i}}\omega$. Making the last term on the right-hand side of Eq. (6.7) the subject, substituting Eqs. (6.8) and (6.9) and changing indices, we get

$$\begin{aligned} \alpha_j X^j &= X^j \frac{\partial \omega_j}{\partial u^i} - \omega_s X^j {\Gamma^s}_{ij} \\ \alpha_j &= \frac{\partial \omega_j}{\partial u^i} - \omega_s {\Gamma^s}_{ij} \end{aligned}$$

[2]This Leibniz rule is stated without proof in this book.

Therefore,

$$D_{\partial_{u^i}}\omega = \left(\frac{\partial \omega_j}{\partial u^i} - \omega_s {\Gamma^s}_{ij}\right) du^j \tag{6.10}$$

We can obtain the covariant derivative of du^i by substituting $du^j = {\delta^j}_k du^k$:

$$\begin{aligned} D_{\partial_{u^i}} du^j &= \left(\frac{\partial {\delta^j}_k}{\partial u^i} - {\delta^j}_s {\Gamma^s}_{ik}\right) du^k \\ &= -{\Gamma^j}_{ik} du^k \end{aligned} \tag{6.11}$$

6.4 The Metric Tensor Is What We Need

The new definition of tangent vectors does not allow the derivation of the induced metric tensor. For intrinsic geometry, we need to assume that the metric tensor is given or derived using another approach.

In the previous chapters, we remarked that many quantities do not use the coordinates in the embedded space $\mathbb{R}^n$, but use u, the metric tensor g_{ij}, the Christoffel symbols ${\Gamma^k}_{ij}$, etc. Examples are the covariant derivative and the geodesic equation. Obviously, all derivatives are done in the space of U. We show shortly that the Christoffel symbols can be written as a function of the metric tensor and its derivatives. Therefore, the metric tensor is more fundamental, and with it, we can perform a lot of calculations.

6.4.1 Tensor product representation of the metric tensor

The metric tensor is a convariant 2-tensor; that is, it takes in two tangent vectors and gives a real number:

$$g : T_uU \times T_uU \mapsto \mathbb{R}$$

We can write the metric tensor as a tensor product of two differential forms:

$$g(\partial_{u^i}, \partial_{u^j}) = g_{st} du^s du^t(\partial_{u^i}, \partial_{u^j}) = g_{st} du^s(\partial_{u^i}) du^t(\partial_{u^j}) = g_{st} {\delta^s}_i {\delta^t}_j = g_{ij}$$

Note that the order of $du^s du^t$ is important when we write the tensor product. In the case of the Riemannian metric, it is symmetric, and hence order is not important.

6.4.2 Covariant derivative of the metric tensor

Lemma 6.4.1 *The covariant derivative of the metric tensor is zero:*

$$D_{\partial_{u^i}} g = 0$$

A proof of the above statement follows from writing the metric tensor as a tensor product of covectors using the Leibniz rule, using Eq. (6.11) and expressing the derivative of the metric tensor in terms of the Christoffel symbols:

$$\begin{aligned} g &= g_{st} du^s du^t \\ D_{\partial_{u^i}} g &= (D_{\partial_{u^i}} g_{st}) du^s du^t + g_{st} (D_{\partial_{u^i}} du^s) du^t + g_{st} du^s (D_{\partial_{u^i}} du^t) \\ D_{\partial_{u^i}} g &= \frac{\partial g_{st}}{\partial u^i} du^s du^t - g_{st} \Gamma^s{}_{ik} du^k du^t - g_{st} \Gamma^t{}_{ik} du^k du^s \end{aligned}$$

Reindexing so that we can factor out the covectors,

$$\begin{aligned} D_{\partial_{u^i}} g &= \frac{\partial g_{st}}{\partial u^i} du^s du^t - g_{kt} \Gamma^k{}_{is} du^s du^t - g_{sk} \Gamma^k{}_{it} du^s du^t \\ &= \left(\frac{\partial g_{st}}{\partial u^i} - g_{kt} \Gamma^k{}_{is} - g_{sk} \Gamma^k{}_{it} \right) du^s du^t \end{aligned}$$

We show that the term in the parenthesis is zero using notation, $\partial_{u^i} = \hat{e}_i$, and use $\partial \hat{e}_i / \partial u^j = \Gamma^k{}_{ij} \hat{e}_k$:

$$\begin{aligned} \frac{\partial g_{st}}{\partial u^i} &= \partial_{u^s} \cdot \frac{\partial \partial_{u^t}}{\partial u^i} + \partial_{u^t} \cdot \frac{\partial \partial_{u^s}}{\partial u^i} \\ &= \partial_{u^s} \cdot \Gamma^k{}_{ti} \partial_{u^k} + \partial_{u^t} \cdot \Gamma^k{}_{si} \partial_{u^k} \\ &= g_{sk} \cdot \Gamma^k{}_{it} + g_{kt} \cdot \Gamma^k{}_{is} \end{aligned}$$

Therefore,

$$D_{\partial_{u^i}} g = 0$$

6.4.3 Christoffel symbols in intrinsic geometry

The Christoffel symbols can be written in terms of the metric tensor:

$$\Gamma^k{}_{ij} = \frac{1}{2} g^{ks} \left(\frac{\partial g_{si}}{\partial u^j} + \frac{\partial g_{sj}}{\partial u^i} - \frac{\partial g_{ij}}{\partial u^s} \right) \tag{6.12}$$

We see that the Christoffel symbols can be derived solely from the metric tensor and coordinates in U. Other quantities that depend on the Christoffel symbols can be calculated.

6.4.4 Normal coordinates

The metric tensor is a function of a coordinate point:

$$g_{ij} = g_{ij}(u)$$

It is symmetric. Indeed, all the values of its entry are non-negative. All real symmetric matrices are diagonalizable. A proof is left as an exercise for the readers. Given a fixed point u_0, we can find a set of linear coordinate transformations such that the metric tensor is diagonal:

$$g(u_0) = O^T \Lambda(u_0) O$$

where O is an orthogonal matrix and d is the diagonal. We consider the case when the space is not compressed so much that the length in a certain direction becomes zero at u_0. Then, none of the eigenvalues (entries of the diagonal) are zero, i.e. $d_{ii} > 0$. Next, we can perform another coordinate transformation $d^{-1/2}$ to make the transformed metric tensor the identity matrix:

$$I = O^T \Lambda^{-1/2} g \Lambda^{-1/2} O$$

Hence, the coordinate transformation that makes the metric tensor into an identity matrix is $\Lambda^{-1/2} O$. We call this coordinate system the "normal coordinate system." This coordinate system is useful when we want to study how the metric tensor changes as we go from one point in the manifold to another. In particular, the Taylor expansion of the metric tensor can be performed, and it is related to the Riemann curvature tensor, which we study in a later chapter of this book.

Exercise 6.4.1

1. Prove Eq. (6.12):

$$\Gamma^k{}_{ij} = \frac{1}{2} g^{ks} \left(\frac{\partial g_{si}}{\partial u^j} + \frac{\partial g_{sj}}{\partial u^i} - \frac{\partial g_{ij}}{\partial u^s} \right) \tag{6.13}$$

Solution: Begin by differentiating the metric tensor:

$$\begin{aligned} \frac{\partial g_{si}}{\partial u^j} &= \frac{\partial \hat{e}_s \cdot \hat{e}_i}{\partial u^j} \\ &= \frac{\partial \hat{e}_s}{\partial u^j} \cdot \hat{e}_i + \hat{e}_s \cdot \frac{\partial \hat{e}_i}{\partial u^j} \\ &= \Gamma^k{}_{sj} \hat{e}_k \cdot \hat{e}_i + \Gamma^k{}_{ij} \hat{e}_s \cdot \hat{e}_k \\ &= \Gamma^k{}_{sj} g_{ki} + \Gamma^k{}_{ij} g_{sk} \end{aligned}$$

Perform the same operations on the other terms of Eq. (6.12):

$$\frac{\partial g_{sj}}{\partial u^i} = \Gamma^k{}_{si} g_{kj} + \Gamma^k{}_{ji} g_{sk} \tag{6.14}$$

$$\frac{\partial g_{ij}}{\partial u^s} = \Gamma^k{}_{is} g_{kj} + \Gamma^k{}_{js} g_{ik}$$

Subtracting,

$$\frac{\partial g_{si}}{\partial u^j} + \frac{\partial g_{sj}}{\partial u^i} - \frac{\partial g_{ij}}{\partial u^s} = 2\Gamma^k{}_{ji} g_{sk}$$

Multiply by the inverse metric tensor g^{ls} on both sides:

$$\begin{aligned}
\Gamma^k{}_{ji} g_{sk} g^{ls} &= \frac{1}{2} g^{ls} \left(\frac{\partial g_{si}}{\partial u^j} + \frac{\partial g_{sj}}{\partial u^i} - \frac{\partial g_{ij}}{\partial u^s} \right) \\
\Gamma^k{}_{ji} \delta^l{}_k &= \frac{1}{2} g^{ls} \left(\frac{\partial g_{si}}{\partial u^j} + \frac{\partial g_{sj}}{\partial u^i} - \frac{\partial g_{ij}}{\partial u^s} \right) \\
\Gamma^l{}_{ji} &= \frac{1}{2} g^{ls} \left(\frac{\partial g_{si}}{\partial u^j} + \frac{\partial g_{sj}}{\partial u^i} - \frac{\partial g_{ij}}{\partial u^s} \right)
\end{aligned}$$

6.5 Pullback and Pushforward

Pushforward and pullback are functions that associate scalar functions and tensors between different manifolds. We first introduce mappings between manifolds and then discuss pushforward.

6.5.1 Maps between manifolds

Consider two manifolds, U and $\hat{U}$. These manifolds do not need to be of the same dimensions. Let φ be a differentiable function which maps

$$\varphi : U \mapsto \hat{U}$$

Note that this map may not have an inverse. Differentiability in the usual calculus way can be defined, for example, as

$$J^i{}_j = \frac{\partial \varphi^i}{\partial u^j}$$

Figure 6.2 illustrates the mappings between manifolds.

Remarks

In extrinsic geometry, we have $\varphi : U \mapsto M$. We can consider U to be a manifold by itself. In this view, φ is a mapping between manifolds.

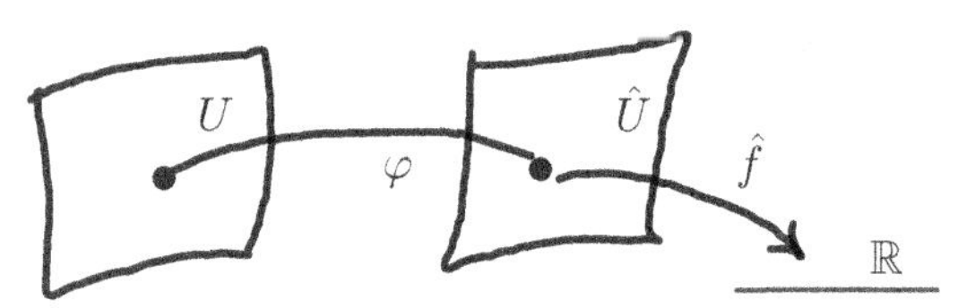

Figure 6.2: Mapping between manifolds U and $\hat{U}$ and a function mapping $\hat{U}$ to $\mathbb{R}$.

6.5.2 Pushforward

Pushforward is a mapping that addresses the issue of how to compare two tangent vectors at different tangent spaces. Given a tangent vector at a point in U, how to move this tangent vector to another point in another manifold $\hat{U}$ while keeping its properties invariant? Copying the tangent components during the translation does not make it invariant. Since tangent vectors are functions on functions, a good way to quantify invariance is that after the translation, its actions on all possible functions remain the same.

First, we establish what we mean by "the same" vector in one manifold. We say two tangent vectors X and Y at $u \in U$ are the same if for all $C^\infty(U)$, the functions $f : U \mapsto \mathbb{R}$, $X(f) = Y(f)$. The differentials are evaluated at the same point in U. Given two manifolds, U and $\hat{U}$, and a function that maps U to $\hat{U}$,

$$\varphi : U \mapsto \hat{U}$$

Taking an arbitrary function $\hat{f} \in C^\infty(\hat{U})$,

$$\hat{f} : \hat{U} \mapsto \mathbb{R}$$

We need to find the function on U which corresponds to $\hat{f}$. This is given by $f = \hat{f} \circ \varphi$:

$$f = \hat{f} \circ \varphi : U \mapsto \mathbb{R}$$

So, we now have two scalar functions f and $\hat{f}$. Given that we have two tangent vectors $X \in T_u U$ and $\hat{X} \in T_{\varphi(u)}\hat{U}$. If $X(\hat{f} \circ \varphi) = \hat{X}(\hat{f})$ for all $\hat{f} \in C^\infty(U)$, then we say that X and $\hat{X}$ are "the same." In terms of differential geometry, we state that

$$\varphi_* X(\hat{f}) = X(\hat{f} \circ \varphi) = \hat{X}(\hat{f})$$

φ_* is called the pushforward operation. Note the mappings:

$$\begin{aligned} \varphi &: U \mapsto \hat{U} \\ \varphi_* &: T_u U \mapsto T_{\varphi(u)}\hat{U} \end{aligned} \tag{6.15}$$

The tangent vector $\hat{X}$ acting on $\hat{f}(\hat{u})$ is

$$\hat{X}(\hat{f}) = \hat{X}^i \partial_{\hat{u}^i} \hat{f}$$

The tangent vector X acting on $f(u)$ is

$$X(f) = X^i \partial_{u^i}(\hat{f} \circ \varphi)$$

Express the partial derivative of u in terms of $\hat{u}$ using the chain rule or similarly using the inverse Jacobian:

$$\partial_{u^i} = \frac{\partial \hat{u}^j}{\partial u^i} \partial_{\hat{u}^j} = \frac{\partial \varphi^j}{\partial u^i} \partial_{\hat{u}^j}$$

Then, the pushforward can be expressed as

$$X^i \partial_{u^i}(\hat{f}(\varphi(u))) = X^i \frac{\partial \varphi^j}{\partial u^i} \partial_{\hat{u}^j}(\hat{f}) \equiv \hat{X}^j \partial_{\hat{u}^j}(\hat{f})$$

Recall that the Jacobian of a transformation from U to $\hat{U}$ is

$$J^j{}_i = \frac{\partial u^j}{\partial \hat{u}^i}$$

Hence, the coefficients of the pushforward vectors are given by

$$\hat{X}^j(\varphi(u)) = \frac{\partial \varphi^j}{\partial u^i} X^i = (J^{-1})^j{}_i X^i(u) \tag{6.16}$$

Definition 6.5.1 (pushforward transformation) Given two manifolds U and $\hat{U}$ and a differentiable map between them φ, and given a vector $X = X^i \partial_{u^i} \in T_u U$, the pushforward of this vector onto $\hat{U}$ via φ is given by the contravariant transformation of its components with X applied to the scalar functions $\hat{f} \circ \varphi$, $\hat{f} : \hat{U} \mapsto \mathbb{R}$. Let the pushforward vector be $\hat{X} = \hat{X}^i \partial_{u^i}$. Then,

$$\hat{X}^j(\varphi(u)) = (J^{-1})_i{}^j X^i(u)$$

$$X(f) = X(\hat{f} \circ \varphi) = \hat{X}(\hat{f})$$

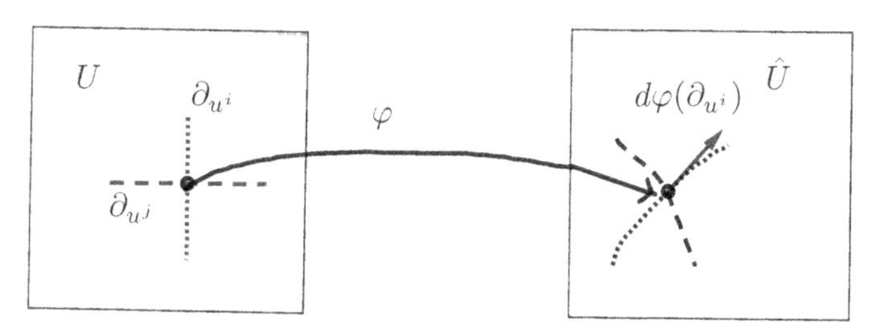

Figure 6.3: Mapping between manifolds U and $\hat{U}$ by φ. Moving along the ∂_{u^i} direction in U, and the image of this line (on $\hat{U}$) is drawn with a dotted line. The dashed line indicates the path along the u^j direction and its image. The arrow indicates the tangent vector along the dotted line $d\varphi(\partial_{u^i})$.

6.5.3 Pushforward and covectors

Writing pushforward in terms of a differential has become a common notation, and it would be good for the reader to become familiar with this notation. For a mapping $\varphi : U \mapsto \hat{U}$, its corresponding covector is given by applying the tangent vector X on each component of φ:

$$\begin{aligned} \varphi &: U \mapsto \hat{U} \\ \varphi^j &: U \mapsto \mathbb{R} \\ d\varphi^j(X) &= X(\varphi^j) \\ d\varphi &: T_u U \mapsto T_{\hat{u}} \hat{U} \\ d\varphi(X) &= X(\varphi^j)\partial_{\hat{u}^j} \end{aligned} \tag{6.17}$$

Note that we inserted a $\partial_{\hat{u}^j}$ into the covector equation to make the output a tangent vector in $T_{\hat{u}}\hat{U}$. The components of tangent vectors, which recover Eq. (6.16), are

$$X(\varphi^j) = X^i(u)\frac{\partial \varphi^j}{\partial u^i} = \hat{X}^j(\varphi(u))$$

Hence,

$$\varphi_* = d\varphi$$

Figure 6.3 illustrates the pushforward of tangent vector ∂_{ui} by $d\varphi$.

6.5.4 Properties of pushforward

Lemma 6.5.1 *Given maps between manifolds $\varphi : U \mapsto \hat{U}$ and $\hat{\varphi} : \hat{U} \mapsto \tilde{U}$. The pushforward between these maps is given by*

$$(\hat{\varphi} \circ \varphi)_* = \hat{\varphi}_* \circ \varphi_*$$

We first state the following identities and definitions:

$$\begin{aligned}
\tilde{f} &: \tilde{U} \mapsto \mathbb{R} \\
\hat{f} &= \tilde{f} \circ \hat{\varphi} \\
f &= \hat{f} \circ \varphi = \tilde{f} \circ \hat{\varphi} \circ \varphi \\
\hat{X}(\hat{f}) &= X(\hat{f} \circ \varphi) \qquad \text{(pushforward, } \hat{X} = \varphi_* X) \\
\tilde{X}(\tilde{f}) &= \hat{X}(\tilde{f} \circ \hat{\varphi}) \qquad \text{(pushforward, } \tilde{X} = \hat{\varphi}_* \hat{X})
\end{aligned} \tag{6.18}$$

Using the above equations and substituting the terms,

$$\begin{aligned}
\tilde{X}(\tilde{f}) &= \hat{X}(\tilde{f} \circ \varphi) \\
&= \hat{X}(\hat{f}) \\
&= X(\hat{f} \circ \varphi) \\
&= X(\tilde{f} \circ \hat{\varphi} \circ \varphi) \\
&= ((\hat{\varphi} \circ \varphi)_* X)(\tilde{f}) \qquad \text{(use the definition of pushforward)}
\end{aligned}$$

We conclude that $\tilde{X} = (\hat{\varphi} \circ \varphi)_* X$. Also, we build $\tilde{X}$ using two pushforward operations, $\tilde{X} = \hat{\varphi}_* \hat{X}$ and $\hat{X} = \varphi_* X$. Combining $\tilde{X} = (\hat{\varphi}_* \circ \varphi_*) X$. Therefore,

$$(\hat{\varphi} \circ \varphi)_* X = (\hat{\varphi}_* \circ \varphi_*) X$$

Since X is arbitrary, $(\hat{\varphi}_* \circ \varphi_*) = (\hat{\varphi} \circ \varphi)_*$.

6.5.5 Pushforward for the case of extrinsic geometry

In extrinsic geometry, we consider the coordinate representation of a manifold M using U. Here, $U \subset \mathbb{R}^d$ and the chart is $\varphi : U \mapsto M$. Now, we consider U in Cartesian coordinates with basis vectors $\hat{e}_i$, and a vector in U is given by

$$X = X^i(u) \hat{e}_i$$

Here, X^i are functions that map u to a real number:

$$X^i : U \mapsto \mathbb{R}$$

We want to ask the question: What is the pushforward of X under φ? With $\varphi = \varphi^j e_j$, using the covector view of pushforward and noting that e_j is the Euclidean basis vector in the embedded space $\mathbb{R}^n$,

$$d\varphi(X) = X(\varphi^j) e_j = X^i \frac{\partial \varphi^j}{\partial u^i} e_j$$

Since e_j are constants and $\varphi = \varphi^j e_j$, we can write

$$d\varphi(X) = X(\varphi^j) e_j = X^i \frac{\partial \varphi}{\partial u^i}$$

Hence,

$$\varphi_* X = d\varphi(X) = X^i(u)\frac{\partial \varphi}{\partial u^i} \tag{6.19}$$

Exercise 6.5.1

1. Consider φ as the mapping from a cylinder to a sphere. Mappings for the sphere and cylinder are given by

$$\begin{aligned} x^1 &= \cos\lambda \\ x^2 &= \sin\lambda \\ x^3 &= z \\ \hat{x}^1 &= \cos\phi\sin\theta \\ \hat{x}^2 &= \sin\phi\sin\theta \\ \hat{x}^3 &= \cos\theta \end{aligned}$$

Note that the extrinsic coordinates φ and $\hat{\varphi}$ are not needed for this exercise since we work in an intrinsic geometry setting. They are given here for the purpose of explaining the question. Let $u^1 = \lambda$, $u^2 = z$, $\hat{u}^1 = \phi$ and $\hat{u}^2 = \theta$. The map φ from the cylinder to the sphere is

$$\begin{aligned} \hat{u}^1 = \phi &= \lambda = u^1 \\ \hat{u}^2 = \theta &= \cos^{-1} z = \cos^{-1} u^2 \end{aligned}$$

Given the tangent vector $X = (X^1, X^2)$:

a. Derive the general form of the pushforward of X under φ.

b. Given the function and the components of the vector X,

$$\begin{aligned} \hat{f}(\theta,\phi) &= \cos^2\phi\sin^2\theta + 2\sin\phi\sin\theta\cos\theta \\ (X^1, X^2) &= (1, \sqrt{1-z^2}) \end{aligned}$$

apply the pushforward vector derived in the previous part of this question on this function. Show that the actions of X on $\hat{f}\circ\varphi$ and the action of $\hat{X}$ on $\hat{f}$ are equal.

Solution:

a. First, observe that the relationship between u and $\hat{u}$ is

$$\begin{aligned} \phi &= \lambda \\ \theta &= \cos^{-1} z \end{aligned}$$

The transformations of the differentials are

$$\begin{aligned}
\frac{\partial \phi}{\partial \lambda} &= 1 \\
\frac{\partial \phi}{\partial z} &= 0 \\
\frac{\partial \theta}{\partial \lambda} &= 0 \\
\frac{\partial \theta}{\partial z} &= -\frac{1}{\sqrt{1-z^2}}
\end{aligned}$$

The components of $\hat{X}$ are

$$\begin{aligned}
\hat{X}^1 &= \frac{\partial \phi}{\partial \lambda} X^1 + \frac{\partial \phi}{\partial z} X^2 = X^1 \\
\hat{X}^2 &= \frac{\partial \theta}{\partial \lambda} X^1 + \frac{\partial \theta}{\partial z} X^2 = -\frac{X^2}{\sqrt{1-z^2}}
\end{aligned} \tag{6.20}$$

The pushforward of X is

$$\hat{X} = X^1 \partial_\phi - \frac{X^2}{\sqrt{1-z^2}} \partial_\theta = \partial_\phi - \partial_\theta \tag{6.21}$$

b. To show $X(\hat{f} \circ \varphi) = \hat{X}(\hat{f})$,

$$\hat{f} = \cos^2 \phi \sin^2 \theta + \sin \phi \sin 2\theta$$

$$(X^1, X^2) = (1, \sqrt{1-z^2})$$

Use Eq. (6.21):

$$\begin{aligned}
\hat{X}(\hat{f}) &= (\partial_\phi - \partial_\theta)\, \hat{f} \\
&= -\sin 2\phi \sin^2 \theta + \cos \phi \sin 2\theta (1 - \cos \phi) - 2 \cos 2\theta \sin \phi
\end{aligned}$$

Write the expression in terms of the u coordinate:

$$\begin{aligned}
\hat{X}(\hat{f}) = \; & -\sin 2\lambda (1 - z^2) + 2z \cos \lambda \sqrt{1-z^2} (1 - \cos \lambda) \\
& + 2(1 - 2z^2) \sin \lambda
\end{aligned}$$

$\hat{f} \circ \varphi$ is, in fact, simply $\hat{f}$ written in terms of λ, z:

$$f(\lambda, z) = (\cos^2 \lambda)(1 - z^2) + 2z \sin \lambda \sqrt{1-z^2}$$

$X(f)$ is then

$$\begin{aligned} X(f) &= \left(\frac{\partial}{\partial\lambda} + \sqrt{1-z^2}\frac{\partial}{\partial z}\right) f \\ &= -\sin 2\lambda(1-z^2) + 2z\cos\lambda\sqrt{1-z^2}(1-\cos\lambda) \\ &\quad + 2\sin\lambda(1-2z^2) \\ &= \hat{X}(\hat{f}) \end{aligned}$$

---

2. Pushforward can be done across a manifold in different dimensions. Consider the map φ as,

$$\begin{aligned} \hat{u}^1 &= \sin u^1 \cos u^2 \\ \hat{u}^2 &= \sin u^1 \sin u^2 \\ \hat{u}^3 &= \cos u^1 \end{aligned}$$

a. Note that this is a map to a 2-sphere. Find the pushforward of vectors onto the sphere for a tangent vector (X^1, X^2).

b. Find the pushforward of the tangent vectors $X = (1, 0)$ and $X = (0, 1)$. Show that these pushforward tangent vectors are indeed the tangent vectors in extrinsic geometry, $\hat{e}_1$ and $\hat{e}_2$.

Solution: First, compute the partials:

$$\begin{aligned} \frac{\partial\varphi^1}{\partial u^1} &= \cos u^1 \cos u^2 \\ \frac{\partial\varphi^1}{\partial u^2} &= -\sin u^1 \sin u^2 \\ \frac{\partial\varphi^2}{\partial u^1} &= \cos u^1 \sin u^2 \\ \frac{\partial\varphi^2}{\partial u^2} &= \sin u^1 \cos u^2 \\ \frac{\partial\varphi^3}{\partial u^1} &= -\sin u^1 \\ \frac{\partial\varphi^3}{\partial u^2} &= 0 \end{aligned}$$

Find the components of $\hat{X}$:

$$\begin{aligned}\hat{X}^1 &= \frac{\partial\varphi^1}{\partial u^1}X^1 + \frac{\partial\varphi^1}{\partial u^2}X^2 = X^1\cos u^1\cos u^2 - X^2\sin u^1\sin u^2\\ \hat{X}^2 &= \frac{\partial\varphi^2}{\partial u^1}X^1 + \frac{\partial\varphi^2}{\partial u^2}X^2 = X^1\cos u^1\sin u^2 + X^2\sin u^1\cos u^2\\ \hat{X}^3 &= \frac{\partial\varphi^3}{\partial u^1}X^1 + \frac{\partial\varphi^3}{\partial u^2}X^2 = -X^1\sin u^1\end{aligned}$$

The basis vectors in the $\hat{U}$ space are

$$\partial_{\hat{u}^1}, \partial_{\hat{u}^2}, \partial_{\hat{u}^3}$$

Therefore, $\hat{X}$ is given by

$$\begin{aligned}\hat{X} &= (X^1\cos u^1\cos u^2 - X^2\sin u^1\sin u^2)\partial_{\hat{u}^1}\\ &\quad + (X^1\cos u^1\sin u^2 + X^2\sin u^1\cos u^2)\partial_{\hat{u}^2} - (X^1\sin u^1)\partial_{\hat{u}^3}\end{aligned}$$

When $X = (1, 0)$, $\hat{X}$ reduces to

$$\hat{X} = \cos u^1\cos u^2\partial_{\hat{u}^1} + \cos u^1\sin u^2\partial_{\hat{u}^2} - \sin u^1\partial_{\hat{u}^3} = \hat{e}_1 \tag{6.22}$$

When $X = (0, 1)$, $\hat{X}$ reduces to

$$\hat{X} = -\sin u^1\sin u^2\partial_{\hat{u}^1} + \sin u^1\cos u^2\partial_{\hat{u}^2} = \hat{e}_2 \tag{6.23}$$

Recall that in extrinsic geometry, $\hat{e}_1$ and $\hat{e}_2$ are given by Eqs. (6.22) and (6.23), respectively. The pushforward tangent vectors exactly match the tangent vectors in extrinsic geometry representations. Note that in extrinsic geometry notations, we do not write the tangent vectors as partials.

6.5.6 Pullback

Similarly to defining mapping between manifolds, we can define a function that maps from a manifold to $\mathbb{R}$:

$$\hat{f} : \hat{U} \mapsto \mathbb{R}$$

Then, the function that maps a point on U onto $\mathbb{R}$ through f is given by

$$f = \hat{f} \circ \varphi = \varphi^* \hat{f}$$

which is called the pullback of $\hat{f}$ by φ. Pullbacks are often denoted with a superscript $*$, as shown above. Differentiability can be defined in the usual calculus sense, e.g. $\partial\hat{f}(\hat{u})/\partial\hat{u}^i$ and

$$\frac{\partial f(u)}{\partial u^j} = \frac{\partial\hat{f}(\varphi(u))}{\partial u^j} = \frac{\partial\hat{f}(\hat{u})}{\partial\hat{u}^i}\frac{\partial\hat{u}^i}{\partial u^j}$$

6.5.7 Pullback of a tensor

Let $\hat{\tau}_{\hat{u}}$ be the tensor field that maps vectors into $\mathbb{R}$:

$$\hat{\tau}_{\hat{u}} : T_{\hat{u}}\hat{U} \times \cdots \times T_{\hat{u}}\hat{U} \mapsto \mathbb{R}$$

$T_{\hat{u}}\hat{U}$ are vector spaces, e.g. tangent vector spaces. Note that this map depends on the point in the manifold u. The pullback of a tensor can be given as follows.

Definition 6.5.2 (pullback of covariant tensors) Let $\hat{\tau}_{\hat{u}}$ be a tensor acting on vectors $X, Y, Z, \ldots$ in the vector space $T_{\hat{u}}\hat{U}$ in $\hat{U}$:

$$\hat{\tau}_{\hat{u}} : T_{\hat{u}}\hat{U} \times \cdots \times T_{\hat{u}}\hat{U} \mapsto \mathbb{R}$$

Given a differentiable map $\varphi : U \mapsto \hat{U}$, with $\hat{u} = \varphi(u)$, the pullback of $\hat{\tau}_{\hat{u}}$ under φ is given by

$$(\varphi^* \hat{\tau})_u(X, Y, Z, \ldots) = \hat{\tau}_{\hat{u}}(\varphi_* X, \varphi_* Y, \varphi_* Z, \ldots)$$

Let's work out what the pullback of a metric tensor is. First, recall that the metric tensor takes in two tangent vectors and produces a real number. Let's first work out the pullback in the extrinsic geometry picture with the chart $\varphi = x$ of the manifold $M \subset \mathbb{R}^n$. Consider the metric tensor in Euclidean space $\mathbb{R}^n$ as the usual dot product of two vectors:

$$g : T_x M \times T_x M \mapsto \mathbb{R}$$

The tangent vectors are given by

$$\hat{e}_i = (\partial_{u^i} \varphi^k) e_k$$

$$\begin{aligned} g(\hat{e}_i, \hat{e}_j) &= (\partial_{u^i} \varphi^k) e_k \cdot (\partial_{u^j} \varphi^l) e_l \\ &= (\partial_{u^i} \varphi^k)(\partial_{u^j} \varphi^l) \delta_{kl} \end{aligned}$$

In intrinsic geometry, the tangent vectors are given by ∂_{u^i}. There is a corresponding metric tensor in the space of U; let this be $\tilde{g}$. We want to show that

$$\tilde{g}(\partial_{u^i}, \partial_{u^j}) = g(\hat{e}_i, \hat{e}_j)$$

This is done through the pullback of g. Using Definition 6.5.2, $\tilde{g} = \varphi^* g$,

$$(\varphi^* g)_u(\partial_{u^i}, \partial_{u^j}) = g_{\varphi(u)}(\varphi_* \partial_{u^i}, \varphi_* \partial_{u^j})$$

Using Eq. (6.17), with $\varphi = x$, we see that

$$\varphi_* \partial_{u^i} = d\varphi(\partial_{u^i}) = (\partial_{u^i} \varphi^k) e_k = \hat{e}_i$$

Hence, the evaluation of the metric tensor in the U space gives

$$\tilde{g}_u(\partial_{u^i}, \partial_{u^j}) = (\varphi^* g)_u(\partial_{u^i}, \partial_{u^j}) = g_{\varphi(u)}(\hat{e}_i, \hat{e}_j)$$

as required.

Exercise 6.5.2

1. Consider covectors in two dimensions,

$$\hat{\omega} = \hat{u}^1\hat{u}^2 d\hat{u}^1 - 2\hat{u}^1 d\hat{u}^2$$

and a mapping φ such that

$$\begin{aligned} \hat{u}^1 &= u^1 \cos u^2 \\ \hat{u}^2 &= u^1 \sin u^2 \end{aligned}$$

Derive an expression for the pullback of $\hat{\omega}$ under φ.

Solution: To get the pullback of a covector, first we need to evaluate the pushforward of tangent vectors:

$$\varphi_* \partial_{u^i} = \partial_{u^i}\varphi^k \partial_{\hat{u}^k}$$

$$\begin{aligned} \varphi_* \partial_{u^1} &= \cos u^2 \partial_{\hat{u}^1} + \sin u^2 \partial_{\hat{u}^2} \\ \varphi_* \partial_{u^2} &= -u^1 \sin u^2 \partial_{\hat{u}^1} + u^1 \cos u^2 \partial_{\hat{u}^2} \end{aligned}$$

$$\begin{aligned} (\varphi^*\hat{\omega})_u(\partial_{u^1}) &= \hat{\omega}_{\varphi(u)}(\cos u^2 \partial_{\hat{u}^1} + \sin u^2 \partial_{\hat{u}^2}) \\ &= (\hat{u}^1\hat{u}^2 d\hat{u}^1 - 2\hat{u}^1 d\hat{u}^2)(\cos u^2 \partial_{\hat{u}^1} + \sin u^2 \partial_{\hat{u}^2}) \\ &= \hat{u}^1\hat{u}^2 \cos u^2 - 2\hat{u}^1 \sin u^2 \\ &= u^1(\cos u^2)u^1 \sin u^2 \cos u^2 - 2u^1 \cos u^2 \sin u^2 \\ \omega^1 &= u^1 \cos u^2 \sin u^2(u^1 \cos u^2 - 2) \end{aligned}$$

$$\begin{aligned} (\varphi^*\hat{\omega})_u(\partial_{u^2}) &= \hat{\omega}_{\varphi(u)}(-u^1 \sin u^2 \partial_{\hat{u}^1} + u^1 \cos u^2 \partial_{\hat{u}^2}) \\ &= (\hat{u}^1\hat{u}^2 d\hat{u}^1 - 2\hat{u}^1 d\hat{u}^2)(-u^1 \sin u^2 \partial_{\hat{u}^1} + u^1 \cos u^2 \partial_{\hat{u}^2}) \\ &= \hat{u}^1\hat{u}^2(-u^1 \sin u^2) - 2\hat{u}^1(u^1 \cos u^2) \\ &= u^1(\cos u^2)u^1 \sin u^2(-u^1 \sin u^2) - 2u^1 \cos u^2(u^1 \cos u^2) \\ \omega^2 &= -(u^1)^2 \cos u^2(u^1 \sin^2 u^2 + 2\cos u^2) \end{aligned}$$

$$\begin{aligned} \omega(u) &= (\varphi^*\hat{\omega})(u) = u^1 \cos u^2 \sin u^2(u^1 \cos u^2 - 2)du^1 \\ &\quad - (u^1)^2 \cos u^2(u^1 \sin^2 u^2 + 2\cos u^2)du^2 \end{aligned}$$

---

2. Define a mapping from the cylinder to the sphere, with the coordinates of the cylinder being

$$\begin{aligned} x^1 &= \cos\lambda \\ x^2 &= \sin\lambda \\ x^3 &= z \end{aligned} \tag{6.24}$$

$z \subset (-1, 1), \phi \subset (0, 2\pi]$. and the coordinates for a 2-sphere being

$$\begin{aligned} \hat{x}^1 &= \cos\phi\sin\theta \\ \hat{x}^2 &= \sin\phi\sin\theta \\ \hat{x}^3 &= \cos\theta \end{aligned} \tag{6.25}$$

$$\begin{aligned} \hat{u}^1 = \theta &= \cos^{-1} z = \cos^{-1} u^1 \\ \hat{u}^2 = \phi &= \lambda = u^2 \end{aligned}$$

Find the pullback of the function $\hat{f} = (\hat{x}^1)^2 + 2\hat{x}^2\hat{x}^3$ in terms of the intrinsic coordinate variables λ, z, θ and ϕ.

Solution: First, we find the mapping between the cylinder and the sphere in terms of intrinsic coordinates:

$$\begin{aligned} \lambda &= \phi \\ z &= \cos\theta \end{aligned}$$

So,

$$\varphi(\lambda, z) = (\theta, \phi) = (\cos^{-1} z, \lambda)$$

$$\hat{f} = \cos^2\phi\sin^2\theta + 2\sin\phi\sin\theta\cos\theta$$

The pullback of $\hat{f}$ under φ is

$$f = \varphi^*\hat{f} = \hat{f}\circ\varphi = \cos^2\lambda(1 - z^2) + 2z\sqrt{1 - z^2}\sin\lambda$$

---

3. Using the map between the cylinder and the sphere in the previous question, find the pullback of the metric tensor. Assume that the metric tensor of the sphere is given by

$$g = \begin{pmatrix} 1 & 0 \\ 0 & \sin^2\theta \end{pmatrix}$$

Solution: First, evaluate the pushforward of the tangent vectors:

$$\begin{aligned}\varphi_*\partial_z &= (\partial_z\theta)\partial_\theta + (\partial_z\phi)\partial_\phi \\ &= -\frac{1}{\sqrt{1-z^2}}\partial_\theta \\ \varphi_*\partial_\lambda &= (\partial_\lambda\theta)\partial_\theta + (\partial_\lambda\phi)\partial_\phi \\ &= \partial_\phi\end{aligned}$$

Now, we are ready to evaluate the pullback:

$$\begin{aligned}(\varphi^* g)_u(\partial_z, \partial_z) &= g(\varphi_*\partial_z, \varphi_*\partial_z) \\ &= \frac{1}{1-z^2} g(\partial_\theta, \partial_\theta) = \frac{1}{1-z^2} \\ (\varphi^* g)_u(\partial_z, \partial_\lambda) &= g(\varphi_*\partial_z, \varphi_*\partial_\lambda) \\ &= -\frac{1}{\sqrt{1-z^2}} g(\partial_\theta, \partial_\phi) = 0 \\ (\varphi^* g)_u(\partial_\lambda, \partial_\lambda) &= g(\varphi_*\partial_\lambda, \varphi_*\partial_\lambda) \\ &= g(\partial_\phi, \partial_\phi) = \sin^2\theta \\ &= (1-\cos^2\theta) = 1 - z^2\end{aligned}$$

Hence,

$$(\varphi^* g) = \begin{pmatrix} \frac{1}{1-z^2} & 0 \\ 0 & 1-z^2 \end{pmatrix}$$

Let's examine the new metric tensor to see if this tensor makes sense: $\lim_{z\to\pm1} \varphi^* g(\partial_z, \partial_z) = \infty$. This means that a small change in z results in a big change in the length measurement on the sphere. Note that near the pole, the sphere is "flat," which means that once we move z slightly, θ changes quite significantly. The limit of the derivative at $z = \pm1$ is infinity. Also, $\lim_{z\to\pm1} \varphi^* g(\partial_\phi, \partial_\phi) = 0$. At the pole, the length measure due to moving along the ϕ direction is zero.

6.6 Summary

Definition 6.2.1 (tangent vectors) Given a manifold U, with $u \in U$, define the basis tangent vector as

$$\hat{e}_i = \frac{\partial}{\partial u^i} = \partial_{u^i}$$

Tangent vectors are given by $X = X^i \partial_{u^i}$, which is a linear mapping

$$X : C^\infty(U) \mapsto \mathbb{R}$$

C^∞ is the set of differentiable functions that maps U to $\mathbb{R}$.

Definition 6.2.2 Tangent vectors can be defined as the derivative of a path on the manifold:

$$\frac{d\gamma(t)}{dt} \equiv \frac{du^i}{dt}\frac{\partial}{\partial u^i} = X^i \partial_{u^i} \tag{6.2}$$

6.6.1 Christoffel symbols

$$\Gamma^k{}_{ij} = \frac{1}{2} g^{ks} \left(\frac{\partial g_{si}}{\partial u^j} + \frac{\partial g_{sj}}{\partial u^i} - \frac{\partial g_{ij}}{\partial u^s} \right)$$

Definition 6.5.1 (pushforward transformation) Given two manifolds U and $\hat{U}$ and a differentiable map between them φ, and given a vector $X = X^i \partial_{u^i} \in T_u U$, the pushforward of this vector onto $\hat{U}$ via φ is given by the contravariant transformation of its components with X applied to the scalar functions $\hat{f} \circ \varphi$, $\hat{f} : \hat{U} \mapsto \mathbb{R}$. Let the pushforward vector be $\hat{X} = \hat{X}^i \partial_{u^i}$. Then,

$$\hat{X}^j(\varphi(u)) = (J^{-1})_i{}^j X^i(u)$$

$$X(f) = X(\hat{f} \circ \varphi) = \hat{X}(\hat{f})$$

6.6.2 Pushforward and covectors

Given the mapping $\varphi : U \mapsto \hat{U}$, the individual component of φ is a map onto $\mathbb{R}$. $\varphi^j : U \mapsto \mathbb{R}$. The covector of φ^j is the pushforward operator

$$\varphi^j_* = d\varphi^j$$

Alternatively, by writing in a general way for all components,

$$\varphi_* = d\varphi$$

Lemma 6.5.1 *Given maps between manifolds $\varphi : U \mapsto \hat{U}$ and $\hat{\varphi} : \hat{U} \mapsto \tilde{U}$. The pushforward between these maps is given by*

$$(\hat{\varphi} \circ \varphi)_* = \hat{\varphi}_* \circ \varphi_*$$

Definition 6.5.2 (pullback of covariant tensors) Let $\hat{\tau}_{\hat{u}}$ be a tensor acting on vectors $X, Y, Z, \ldots$ in the vector space $T_{\hat{u}}\hat{U}$ in $\hat{U}$:

$$\hat{\tau}_{\hat{u}} : T_{\hat{u}}\hat{U} \times \cdots \times T_{\hat{u}}\hat{U} \mapsto \mathbb{R}$$

Given a differentiable map $\varphi : U \mapsto \hat{U}$, with $\hat{u} = \varphi(u)$, the pullback of $\hat{\tau}_{\hat{u}}$ under φ is given by

$$(\varphi^* \hat{\tau})_u (X, Y, Z, \ldots) = \hat{\tau}_{\hat{u}}(\varphi_* X, \varphi_* Y, \varphi_* Z, \ldots)$$

Chapter 7

The Lie Derivative

From this chapter onward, we consider intrinsic geometry formalism unless otherwise stated. We no longer use the symbol M to represent a manifold but use the symbol U instead. The key concepts we introduce in this chapter are flow fields and derivatives of one vector field in the direction of another vector field (the Lie derivative).

7.1 Flow Fields

Given a differentiable tangent vector field X, we have a vector at every point on U. If at each point, we "follow" the direction of the tangent vector at this point and take a small step, we reach a neighboring point. From this neighboring point, we again "follow" the direction of the tangent vector at this new point and take a small step. If we keep doing this, we can trace out a path γ in U. Furthermore, the direction of the path will always be tangent to the vector field. Figure 7.1 illustrates the two paths on a vector field. This process can be summarized by

$$\begin{aligned} \gamma(0) &= u_0 \\ \dot{\gamma}(t) &= X(t) \end{aligned} \tag{7.1}$$

u_0 is the starting point of the path. We call the path traced out by the vector field the flow or the integral curve. Figure 7.1 shows one path traced out by X. Since flow fields are special paths, we denote them by

$$\sigma_{X,t}(u_0) = \gamma(t) \text{ s.t. } \gamma(0) = u_0, \dot{\gamma}(t) = X(t) \tag{7.2}$$

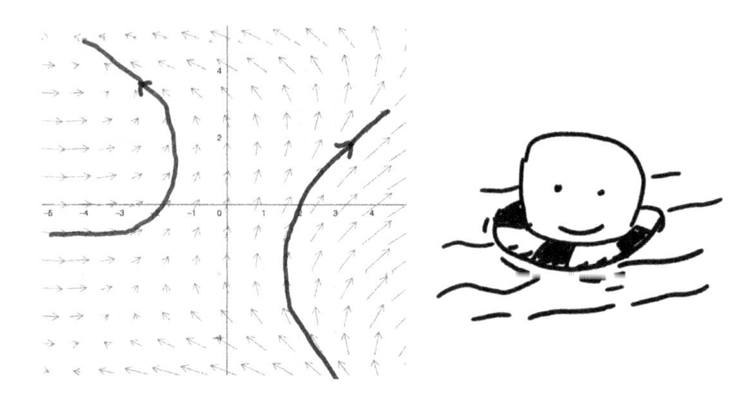

Figure 7.1: Given a vector field, paths can be traced out by following the directions of the vectors at each point.

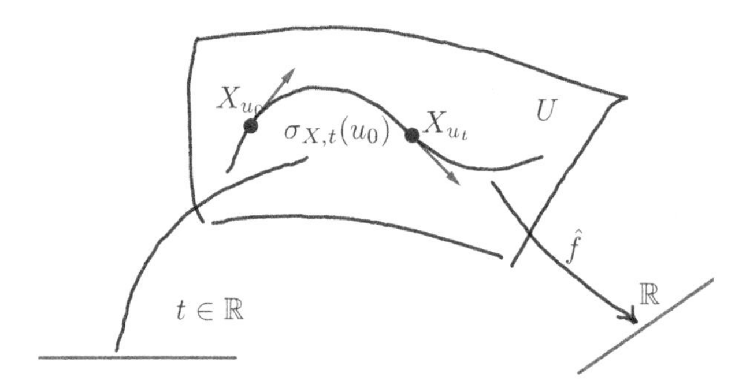

Figure 7.2: Starting from u_0, a flow path is generated for time t to reach the point u_t. $X(u_0)$ and $X(u_t)$ are tangent vectors of the vector field X on the flow path.

$$\begin{aligned} \frac{d\gamma(t)}{dt} &= \frac{du^i}{dt}\partial_{u^j} \\ &= \dot{\gamma}^j(t)\partial_{u^j} \end{aligned}$$

We can associate $\gamma^j(t) \equiv u^j(t)$. Write the flow as

$$\sigma_{X,t}(u_0) = \gamma(t)$$

Figure 7.2 illustrates two tangent vectors X_{u_0} and X_{u_t} along a flow path generated by $\sigma_{x,t}(u_0)$.

7.1.1 Properties of flow fields

Flow fields have interesting properties that lead to important results in the study of Lie derivatives and curvatures:

1. Closure: two successive applications of flow for times t_0 and t_1 are equivalent to a single flow with time $t_0 + t_1$. Furthermore, flows commute:

$$\sigma_{X,t_0+t_1} = \sigma_{X,t_1} \circ \sigma_{X,t_0} = \sigma_{X,t_0} \circ \sigma_{X,t_1}$$

7.1.2 Pushforward and flows

Theorem 7.1.1 *Let X be a smooth vector field on U. Then, the flow path generated by X starting from u_0 and the pushforward of X under $\sigma_{X,t}$ satisfy,*[1]

$$\sigma_{X,t}(u_0)_* X_{u_0} = X_{\sigma_{X,t}(u_0)} \tag{7.3}$$

To prove the above theorem, we consider two flow paths. As shown in Figure 7.3, the first flow path is generated using $\sigma_{X,t}(u_0)$ and the second flow path is generated first by the flow of infinitesimal distance Δt to $u_{\Delta t}$ and then by $\sigma_{X,t}(u_{\Delta t})$. Two end points are mapped by an arbitrary function $\hat{f}$ to λ_t and $\lambda_{t+\Delta t}$, respectively. Note that since these paths are generated by flows, the tangent vector at u_t is given by

$$X_{u_t}(\hat{f}) = \lim_{\Delta t \to 0} \frac{\lambda_{t+\Delta t} - \lambda_t}{\Delta t}$$

However, we know that λ's are also given by the composite function $\lambda_t = \hat{f} \circ \sigma_{X,t}(u_0)$. So,

$$X_{u_0}(\hat{f} \circ \sigma_{X,t}) = \lim_{\Delta t \to 0} \frac{\lambda_{t+\Delta t} - \lambda_t}{\Delta t}$$

Therefore,

$$X_{u_0}(\hat{f} \circ \sigma_{X,t}) = X_{u_t}(\hat{f}) = \sigma_{X,t_*} X_{u_0}(\hat{f})$$

Since $\hat{f}$ is arbitrary, we can write

$$\sigma_{X,t_*} X_{u_0} = X_{u_t}$$

Theorem 7.1.2 *Let X be a vector field in U with flow $\sigma_{X,t}$ and $\hat{X}$ be a vector field in $\hat{U}$ with flow $\sigma_{\hat{X},t}$. Let $\varphi : U \mapsto \hat{U}$. Then, φ_* relates to the flows by*[2]

$$\sigma_{\hat{X},t} \circ \varphi = \varphi \circ \sigma_{X,t} \iff \varphi_* X = \hat{X}. \tag{7.4}$$

---

[1]See John Lee, *Introduction to Smooth Manifolds*, p. 442 [8].
[2]See John Lee, *Introduction to Smooth Manifolds*, p. 468 [8].

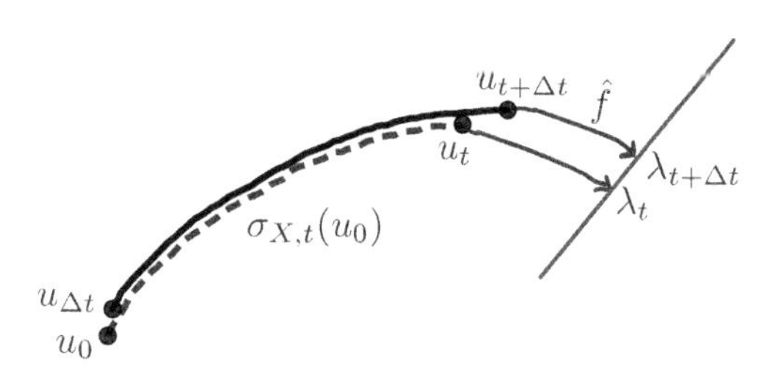

Figure 7.3: The point u_0 flows a little bit by $\sigma_{X,\Delta t}(u_0)$ to $u_{\Delta t}$ and u_0 also flows by a finite t by $\sigma_{X,t}(u_0)$ to u_t. $u_{\Delta t}$ again flows by $\sigma_{X,t}(u_{\Delta t})$ to $u_{t+\Delta t}$. At the points u_t and $u_{t+\Delta t}$, we use an arbitrary function $\hat{f}$ to map to $\mathbb{R}$, yielding λ_t and $\lambda_{t+\Delta t}$.

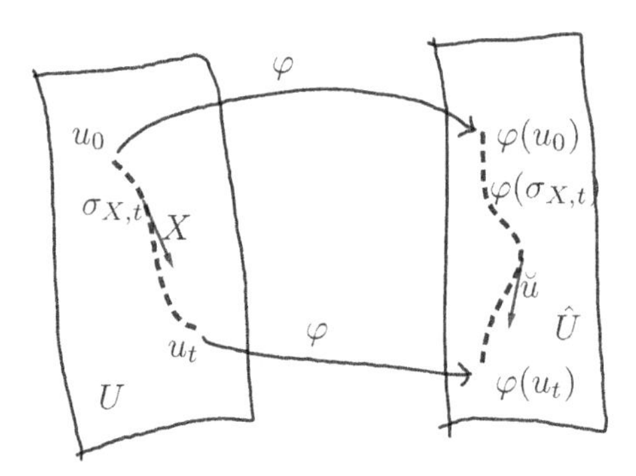

Figure 7.4: Given the mapping between two manifolds, $\varphi : U \mapsto \hat{U}$, and two vector fields, X and $\hat{X}$, one could compose the composite mappings, $\sigma_{\hat{X},t} \circ \varphi$ and $\varphi \circ \sigma_{X,t}$. They are equal iff the two vector fields are related by pushforwards, $\varphi_* X = \hat{X}$.

Figure 7.4 illustrates that φ maps the flow path (integral curve) $\sigma_{X,t}$ at U to the path $\gamma(t)$ at $\hat{U}$. Suppose $\varphi_* X = \hat{X}$, and we want to prove

$$\varphi(\sigma_{X,t}(u)) = \sigma_{\hat{X},t}(\varphi(u))$$

The pushforward equation is

$$\hat{X}^k = \frac{\partial \varphi^k}{\partial u^j} X^j$$

Consider infinitesimal flow $t \ll 1$. Then, the flow at U is

$$\sigma_{X,t}(u) = u + Xt + O(t^2)$$

The corresponding flow in U' written in components is

$$\sigma^k_{\hat{X},t}(\varphi(u)) = \varphi^k(u) + \hat{X}^k t + O(t^2)$$

Now, we map the flow in U onto U':

$$\begin{aligned} \varphi^k(\sigma_{X,t}(u)) &= \varphi^k(u + Xt + O(t^2)) \\ &= \varphi^k(u) + \frac{\partial \varphi^k}{\partial u^j} X^k t + O(t^2) \end{aligned}$$

Since $\hat{X}^k = (\partial\varphi^k/\partial u^j)X^j$, we have

$$\begin{aligned} \varphi^k(\sigma_{X,t}(u)) &= \varphi^k(u) + \hat{X}^k t + O(t^2) \\ &= \sigma^k_{\hat{X},t}(\varphi(u)) \end{aligned}$$

Next, we need to prove the converse, assuming $\varphi(\sigma_{X,t}(u)) = \sigma_{\hat{X},t}(\varphi(u))$. Using Taylor series expansion on both sides,

$$\begin{aligned} \varphi(u + Xt + O(t^2)) &= \varphi(u) + \hat{X}t + O(t^2) \\ \varphi(u) + \frac{\partial \varphi}{\partial u^j} X^j t &= \varphi(u) + \hat{X}t + O(t^2) \end{aligned}$$

We cancel terms on both sides and disregard higher-order terms in t:

$$\frac{\partial \varphi}{\partial u^j} X^j = \hat{X}$$

which is the pushforward equation. Therefore, $\varphi_* X = \hat{X}$. The above proof is valid for infinitesimal flows. We argue that a finite flow is made up of many infinitesimal flows. Therefore, we have Theorem 7.1.2.

Exercise 7.1.1

1. Consider the vector field in $\mathbb{R}^2$:

$$X = -x^2 e_1 + (x^1 - 2x^2)e_2$$

   a. Draw the vector field.

   b. Derive the flow field starting from $(x^1, x^2) = (5, 0)$.

Solution: Using the flow equation (Eq. (7.1)),

$$\begin{aligned} \frac{d\gamma^1}{dt} &= -\gamma^2 \\ \frac{d\gamma^2}{dt} &= \gamma^1 - 2\gamma^2 \end{aligned} \tag{7.5}$$

Differentiate again:

$$\frac{d^2\gamma^2}{dt^2} + 2\frac{d\gamma^2}{dt} + \gamma^2 = 0 \tag{7.6}$$

This is an equation for a damped harmonic oscillator; let's solve it. First, we use the exponential *ansatz* by assuming the solutions of the form

$$\begin{aligned} \gamma_1^2(t) &= C_1 \exp(\lambda t) \\ \gamma_2^2(t) &= C_2 t \exp(\lambda t) \end{aligned}$$

Using the first equation to find λ,

$$\begin{aligned} \dot{\gamma}_1^2(t) &= \lambda C_1 \exp(\lambda t) \\ \ddot{\gamma}_1^2(t) &= \lambda^2 C_1 \exp(\lambda t) \end{aligned}$$

Substituting back into Eq. (7.6),

$$\lambda^2 + 2\lambda + 1 = (\lambda + 1)^2 = 0$$

This implies that $\lambda = -1$. So,

$$\gamma^2(t) = C_1 \exp(-t) + C_2 t \exp(-t)$$

Using $\gamma^2(0) = 0$, we get $C_1 = 0$. Therefore,

$$\begin{aligned} \gamma^2(t) &= C_2 t \exp(-t) \\ \dot{\gamma}^2(t) &= C_2 \exp(-t)(1 - t) \end{aligned}$$

Substitute into Eq. (7.5) to get γ^{q_x},

$$\begin{aligned} \gamma^1(t) &= \dot{\gamma}^2(t) + 2\gamma^2(t) \\ &= C_2 \exp(-t)(1 + t) \end{aligned}$$

Let $\gamma^1(0) = 5$, $C_2 = 5$. The final solution is

$$\gamma(t) = 5 \exp(-t)(1 + t, t)$$

Solving the flow equations with an arbitrary start point (a_x, a_y),

$$\begin{aligned} \gamma^1(t) &= \exp(-t)(C_1 + C_2(t + 1)) \\ \gamma^2(t) &= \exp(-t)(C_1 + C_2 t) \end{aligned}$$

$\gamma^2(0) = a_2$, $C_1 = a_2$, $\gamma^1(0) = a_1 \implies C_2 = a_1 - a_2$. Finally, we can write the flow field as

$$\sigma_{X,t}((a_1, a_2)) = \exp(-t)[a_1 + (a_1 - a_2)t, a_2 + (a_1 - a_2)t] \quad (7.7)$$

How do we check this solution? For a flow, $\dot{\gamma}(t) = X(t)$. We shall check if this holds:

$$\begin{aligned}
\gamma^1(t) &= \exp(-t)[a_1 + (a_1 - a_2)t] \\
\dot{\gamma}^1(t) &= -\gamma^1(t) + \exp(-t)(a_1 - a_2) = -\gamma^{q_2} \\
\gamma^2(t) &= \exp(-t)[a_2 + (a_1 - a_2)t] \\
\dot{\gamma}^2(t) &= -\gamma^2(t) + \exp(-t)(a_1 - a_2) = \gamma^{q_1} - 2\gamma^{q_2}
\end{aligned}$$

We denote the path by γ, which specifies the location of the point along the path, that is, $\gamma^1 = x^1$ and $\gamma^2 = x^2$. As expected, we get

$$X = (\dot{\gamma}^1, \dot{\gamma}^2) = -x^2 e_1 + (q_1 - 2q_2)e_2$$

---

2. Consider a vector field in the 2-sphere, parameterized by $u^1 = \theta$ and $u^2 = \phi$,

$$X = -\phi\partial_\theta + (\theta - 2\phi)\partial_\phi$$

and the flow starting from $u_0 = (\theta, \phi)$. Let $\sigma_{X,t}(u_0)$ be the flow field such that, $u_t = \sigma_{X,t}(u_0)$.

a. Derive the pushforward of X at u_t.

b. Given the function and initial vector X and flow by $t = \ln 2$,

$$\begin{aligned}
\hat{f}(\hat{\theta}, \hat{\phi}) &= \frac{1}{2}\hat{\theta}^2 + \hat{\phi} \\
(X^1, X^2) &= (1, 1)
\end{aligned}$$

apply the pushforward vector derived in the previous part of this question on this function via a flow of $t = \ln 2$. Show that the derivative of X on the pullback of $\hat{f}$ and the derivative of $\hat{X}$ on $\hat{f}$ are equal.

Solution:

a. The flow equation from Eq. (7.7) is

$$\begin{aligned}
\sigma_{X,t}((\theta, \phi)) &= \exp(-t)\,[\theta + (\theta - \phi)t, \phi + (\theta - \phi)t] \quad (7.8) \\
\hat{\theta} &= \exp(-t)[\theta + (\theta - \phi)t] \\
\hat{\phi} &= \exp(-t)[\phi + (\theta - \phi)t]
\end{aligned}$$

so that

$$\begin{aligned}\frac{\partial\hat{\theta}}{\partial\theta} &= \exp(-t)[1+t] \\ \frac{\partial\hat{\theta}}{\partial\phi} &= \exp(-t)[-t] \\ \frac{\partial\hat{\phi}}{\partial\theta} &= \exp(-t)[t] \\ \frac{\partial\hat{\phi}}{\partial\phi} &= \exp(-t)[1-t]\end{aligned} \tag{7.9}$$

The Jacobian is

$$J^{-1} = \exp(-t)\begin{pmatrix} 1+t & -t \\ t & 1-t \end{pmatrix}$$

The tangent basis vectors transform as

$$\begin{aligned}\partial_\theta &= \exp(-t)\left((1+t)\partial_{\hat{\theta}} + t\partial_{\hat{\phi}}\right) \\ \partial_\phi &= \exp(-t)\left(-t\partial_{\hat{\theta}} + (1-t)\partial_{\hat{\phi}}\right)\end{aligned} \tag{7.10}$$

Given the components of X as X^1, X^2, the pushforward of X is

$$\begin{aligned}\hat{X} &= \exp(-t)\left((X^1(1+t) - X^2 t)\partial_{\hat{\theta}} + (X^1 t + X^2(1-t))\partial_{\hat{\phi}}\right) \\ &= \exp(-t)\left((X^1 + (X^1 - X^2)t)\partial_{\hat{\theta}} + (X^2 + (X^1 - X^2)t)\partial_{\hat{\phi}}\right)\end{aligned}$$

b. The coordinate transformation via a flow of $t = \ln 2$ is

$$\begin{aligned}\sigma_{X,\ln 2}((\theta,\phi)) &= \exp(-\ln 2)\,[\theta + (\theta-\phi)\ln 2, \phi + (\theta-\phi)\ln 2] \\ \hat{\theta} &= \frac{\theta(1+\ln 2)}{2} - \frac{\phi\ln 2}{2} \\ \hat{\phi} &= \frac{\theta\ln 2}{2} + \frac{\phi(1-\ln 2)}{2}\end{aligned}$$

The pushforward vector $\hat{w}$ is given by

$$\hat{X}(\hat{f}) = \frac{1}{2}\frac{\partial}{\partial\hat{\theta}} + \frac{1}{2}\frac{\partial}{\partial\hat{\phi}}$$

Applying to $\hat{f} = \hat{\theta}/2 + \hat{\phi}$ gives

$$\begin{aligned}\hat{X}(\hat{f}) &= \frac{1}{2}\frac{\partial(\hat{\theta}^2/2 + \hat{\phi})}{\partial\hat{\theta}} + \frac{1}{2}\frac{\partial(\hat{\theta}^2/2 + \hat{\phi})}{\partial\hat{\phi}} \\ &= \frac{1}{2}\hat{\theta} + \frac{1}{2}\end{aligned}$$

Express $\hat{f}$ in the u coordinates:

$$f(\theta,\phi) = \frac{1}{2}\left(\frac{\theta(1+\ln 2)}{2} - \frac{\phi\ln 2}{2}\right)^2 + \frac{\theta\ln 2}{2} + \frac{\phi(1-\ln 2)}{2}$$

$X(f)$ is,

$$\begin{aligned} X(f) &= \left(\frac{\partial}{\partial\theta} + \frac{\partial}{\partial\phi}\right) f(\theta,\phi) \\ &= \frac{1}{2}\hat{\theta} + \frac{1}{2} = \hat{X}(\hat{f}) \end{aligned}$$

---

3. Consider the vector field in the 2-sphere,

$$X = -\phi\partial_\theta + (\theta - 2\phi)\partial_\phi$$

a. Draw the vector field.

b. Derive the flow field starting from (θ_0, ϕ_0).

Note that we have used the coordinate representation θ, ϕ, and hence the flow field will be derived in terms of θ, ϕ.

Solution: The differential equation is the same as for the flow equation in $\mathbb{R}^2$, with the starting points θ_0, ϕ_0:

$$\sigma_{X,t}((\theta_0, \phi_0)) = \exp(-t)[\theta_0 + (\theta_0 - \phi_0)t, \phi_0 + (\theta_0 - \phi_0)t] \quad (7.11)$$

This gives the flow in the coordinates θ, ϕ. We can map this flow onto the 2-sphere in $\mathbb{R}^3$ using the chart of the 2-sphere by θ, ϕ:

$$\begin{aligned} x^1 &= \cos\phi(t)\sin\theta(t) \\ x^2 &= \sin\phi(t)\sin\theta(t) \\ x^3 &= \cos\theta(t) \end{aligned}$$

$$\begin{aligned} \theta(t) &= \exp(-t)(\theta_0 + (\theta_0 - \phi_0)t) \\ \phi(t) &= \exp(-t)(\phi_0 + (\theta_0 - \phi_0)t) \end{aligned}$$

The reader can use a parametric plotting software with the above equations to plot out the path.

7.2 The Lie Derivative of a Vector

After understanding pullbacks and pushforwards, we are now ready to understand the Lie derivative. Our objective is to find the change in a vector field Y along the flow field of another vector field X.

How do we compute the direction derivative of one vector field in the direction of another vector field? We shall motivate the solution of this question with several failed attempts before presenting the generally accepted solution of the Lie derivative. Note that it is important to discuss why certain approaches do not work in order to understand the real reason for doing things that do work:

1. Since X is a tangent vector field, we could compute the flow field of X starting from the point u_0, i.e. $\sigma_{X,t}(u_0)$. Then, define the derivative as
$$\partial_X Y = \lim_{\epsilon \to 0} \frac{Y(\sigma_{X,\epsilon}(u_0)) - Y(u_0)}{\epsilon}$$
Note that u_0 and $\sigma_{X,\epsilon}(u_0)$ are different points in U. Therefore, $Y(u_0) \in T_{u_0}U$ and $Y(\sigma_{X,\epsilon}(u_0)) \in T_{\sigma_{X,\epsilon}(u_0)}U$ belong to different tangent spaces. Subtracting one tangent vector from another tangent vector in different spaces makes no sense.

2. What if we can move the flowed vector $Y(\sigma_{X,\epsilon}(u_0))$ from the point $\sigma_{X,\epsilon}(u_0)$ to u_0 in some reasonable way? Certainly, since we learned about parallel transport, would parallel transporting Y from the point $\sigma_{X,\epsilon}(u_0)$ to u_0 be a valid approach? It is indeed a good question to ask: Why do we not just use parallel transport to move the vector infinitesimally?

3. Let $Y_\epsilon(u_0)$ be the tangent vector being moved from $\sigma_{X,\epsilon}(u_0)$ to u_0. If we can find an acceptable way without invoking parallel transport, then we get to define a derivative. This is called the Lie derivative.

We have laid down the foundations of taking a derivative properly on U; the rest of the work is to figure out how to move a vector along a flow field effectively. For this, we need some more theoretical foundations, which we provide in the following few sections. We then return to formulating the Lie derivative.

The idea of the Lie derivative is to take an infinitesimal flow ϵ using the vector field X, perform a pushforward of the other vector field Y to the same point in the manifold, take the difference in Y and then take the limit $\epsilon \to 0$. Finally, the Lie derivative applied to a function f is

$$\mathcal{L}_X Y = \lim_{\epsilon \to 0} \frac{(\sigma_{X,-\epsilon})_* Y(u_\epsilon) - Y(u_0)}{\epsilon} \tag{7.12}$$

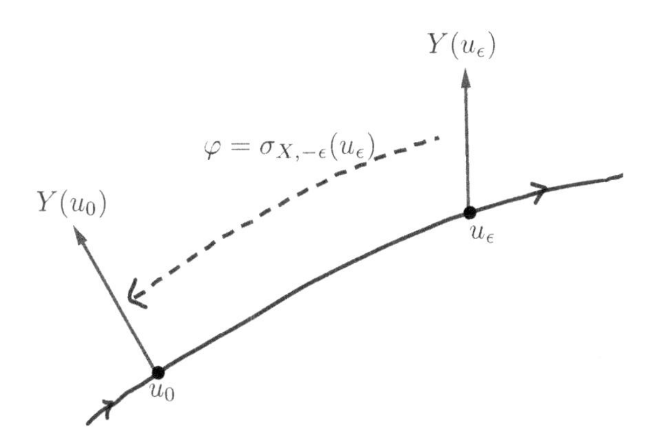

Figure 7.5: To perform the Lie derivative, the tangent vector $Y(u_\epsilon)$ is pushed forward from u_ϵ to u_0 by $\sigma_{X,-\epsilon}(u_\epsilon)$.

Figure 7.5 illustrates the flow field $\sigma_{X,\epsilon}$ and the pushforward of Y from u_ϵ to u_0. We derive the Lie derivative using coordinates in intrinsic geometry. This will facilitate the differentiation of functions and the expansion of functions using the Taylor series. The vector fields are given as

$$\begin{aligned} X &= X^i(u)\partial_{u^i} \\ Y &= Y^i(u)\partial_{u^i}. \end{aligned}$$

7.2.1 Evaluation of pushforward

Express the flow field generated by X in terms of coordinates, $\sigma_{X,\epsilon}(u_0)$. With infinitesimal flow ϵ,

$$u_\epsilon = \sigma_{X,\epsilon}(u_0) \tag{7.13}$$

Similarly,

$$u_0 = \sigma_{X,-\epsilon}(u_\epsilon) \tag{7.14}$$

From Eq. (7.2), we see that

$$\left.\frac{\partial \sigma^i_{X,-\epsilon}}{\partial \epsilon}\right|_{\epsilon=0,u_\epsilon} = X^i(u_\epsilon) = X^i_\epsilon \tag{7.15}$$

We want to express the pushforward from u_ϵ to u_0. Hence, we consider the mapping between manifolds φ as $\varphi(u_\epsilon) = \sigma_{X,-\epsilon}(u_\epsilon)$. Let $Y = Y^i\partial_{u^i_\epsilon}$.

The pushforward equation was given previously by

$$\hat{Y} = Y^i(u_\epsilon)\frac{\partial \sigma^j_{X,-\epsilon}(u_\epsilon)}{\partial u^i_\epsilon}\partial_{u^j} \tag{7.16}$$

We evaluate term by term in Eq. (7.24) and then assemble them. Expand $\sigma_{X,-\epsilon}$ about ϵ to the first order, $\sigma^i_{X,-\epsilon}(u_\epsilon) \approx \sigma^i_{X,0}(u_\epsilon) - \epsilon\partial\sigma^i_{X,0}/\partial\epsilon$,

$$\frac{\partial \sigma^j_{X,-\epsilon}(u_\epsilon)}{\partial u^j_\epsilon} \approx \frac{\partial}{\partial u^j_\epsilon}\left(\sigma^i_{X,0}(u_\epsilon) - \epsilon \left.\frac{d\sigma^i_{X,-\epsilon}}{d\epsilon}\right|_{\epsilon=0,u_\epsilon}\right) \tag{7.17}$$

We need to understand what the derivative of $\sigma_{X,\epsilon}$ is when evaluated at $\epsilon = 0$. Using Eq. (7.15), noting that $\sigma^i_{X,0}(u_\epsilon) = u^i_\epsilon$ and expanding to the first order of ϵ,

$$\frac{\partial \sigma^j_{X,-\epsilon}(u_\epsilon)}{\partial u^j_\epsilon} = \frac{\partial}{\partial u^j_\epsilon}\left(u^i_\epsilon - \epsilon X^i_\epsilon\right) = \delta^i{}_j - \epsilon\frac{\partial X^i_\epsilon}{\partial u^j_\epsilon} + O(\epsilon^2) \tag{7.18}$$

Next, we need to express $Y^j(u_\epsilon)$ as a first-order expansion around the point u_0:

$$\begin{aligned} Y^j_\epsilon = Y^j(u_\epsilon) &= Y^j(\sigma_{X,\epsilon}(u_0)) \qquad (7.19)\\ &\approx Y^j\left(\sigma_{X,0}(u_0) + \epsilon\frac{d\sigma_{X,\epsilon}}{d\epsilon}\right)\\ &\approx Y^j(u_0 + \epsilon X_0)\\ &\approx Y^j(u_0) + \epsilon X^k_0 \left.\frac{\partial Y^j}{\partial u^k}\right|_{u_0} \qquad \text{(Taylor expansion of } Y^j\text{)}\\ &\approx Y^j_0 + \epsilon X^k_0\frac{\partial Y^j_0}{\partial u^k_0} \end{aligned}$$

Substitute Eqs. (7.18) and (7.19) into Eq. (7.24): equation,

$$\begin{aligned} \hat{Y} &\approx \left(Y^j_0 + \epsilon X^k_0\frac{\partial Y^j_0}{\partial u^k_0}\right)\left(\delta^i{}_j - \epsilon\frac{\partial X^i_\epsilon}{\partial u^j_\epsilon}\right)\frac{\partial}{\partial u^i_0} \qquad (7.20)\\ &\approx Y^i_0\frac{\partial}{\partial u^i_0} + \left(\epsilon X^k_0\frac{\partial Y^i_0}{\partial u^k_0} - \epsilon Y^j_0\frac{\partial X^i_\epsilon}{\partial u^j_\epsilon}\right)\frac{\partial}{\partial u^i_0} + O(\epsilon^2) \end{aligned}$$

The term $\epsilon Y^j_0 \partial X^i_\epsilon/\partial u^j_\epsilon$ is first order in ϵ. Expanding this term by replacing u_ϵ by u_0 will generate terms of $O(\epsilon^2)$, and hence we just keep to the first order ϵ and write the expression as $\epsilon Y_0^j \partial X_0^i / \partial u_0^j$. We also change the dummy indices $k \to j$:

$$\hat{Y} = (\sigma_{X,-\epsilon})_* Y(u_\epsilon) = Y(u_0) + \left(\epsilon X_0^j \frac{\partial Y_0^i}{\partial u_0^j} - \epsilon Y_0^j \frac{\partial X_0^i}{\partial u_0^j} \right) \frac{\partial}{\partial u_0^i} + O(\epsilon^2) \quad (7.21)$$

Using Eq. (7.12),

$$\mathcal{L}_X Y = \lim_{\epsilon \to 0} \frac{(\sigma_{X,-\epsilon})_* Y(u_\epsilon) - Y(u_0)}{\epsilon} = \lim_{\epsilon \to 0} \frac{\hat{Y}(u_0) - Y(u_0)}{\epsilon}$$

We get

$$\mathcal{L}_X Y = \left(X^j \frac{\partial Y^i}{\partial u^j} - Y^j \frac{\partial X^i}{\partial u^j} \right) \partial_{u^i} \Big|_{u_0} \quad (7.22)$$

For clarity, the subscript 0 is dropped and replaced by the evaluation of differentials at u_0. We can see that the Lie derivative is also a tangent vector, $\mathcal{L}_X Y \in T_u U$, because it is of the form $\lambda^i \partial_{u^i}$, where λ^i is the first term in the above equation.

7.3 Lie Derivative of Tensors

For the Lie derivative of a vector, we define the difference between the pushforward, in the negative flow direction, of a vector with respect to the flow field of another vector. It seems somewhat awkward to define pushforward in a negative flow direction. The reason is that, as far as I understand, the pullback of a vector is not well defined.

In this section, we study the Lie derivative of a covariant tensor, which is an object that takes in vectors and produces a scalar. For this type of Lie derivative, we cannot use the pushforward trick we used for vectors because the pushforward of a covariant tensor is not well defined. Instead, we compute the Lie derivative by the pullback of a tensor along a flow field of another vector. Instead of deriving the general expression for a tensor, we derive the Lie derivative for a covector. There will be fewer indices, and the derivation will become easier to understand. Let ω be a covector. Then,

$$\mathcal{L}_X \omega = \lim_{\epsilon \to 0} \frac{\sigma^*_{X,\epsilon} \omega_{u_\epsilon} - \omega_{u_0}}{\epsilon} \quad (7.23)$$

Start from u_0, flow by epsilon $\sigma_{X,\epsilon}$ and then pull that tensor field back. We have to switch to using the subscript notation for the evaluation of the covector at a point in the manifold. This is because we need to apply this covector on a vector.

7.3.1 Pullback of covectors

Using a similar trick, we express the pullback of the covector in terms of the first-order Taylor series expansion of ϵ. The pullback of ω is given by

$$\sigma^*_{X,\epsilon}\omega(Y) = \omega_{\sigma_{X,\epsilon}}((\sigma_{X,\epsilon})_* Y)$$

From Eq. (7.24),

$$(\sigma_{X,\epsilon})_* Y_0 = Y^i_\epsilon \frac{\partial}{\partial u^i_\epsilon} - \epsilon \left(X^j_\epsilon \frac{\partial Y^i_\epsilon}{\partial u^j_\epsilon} - Y^j_\epsilon \frac{\partial X^i_\epsilon}{\partial u^j_\epsilon} \right) \frac{\partial}{\partial u^i_\epsilon} + O(\epsilon^2) \tag{7.24}$$

Write ω in terms of its components, $\omega = \omega_k du^k$, and then apply to the pushforward vector:

$$\begin{aligned} du^k((\sigma_{X,\epsilon})_* Y_0) &= du^k(Y_\epsilon) - \epsilon \left(X^j_\epsilon \frac{\partial Y^i_\epsilon}{\partial u^j_\epsilon} - Y^j_\epsilon \frac{\partial X^i_\epsilon}{\partial u^j_\epsilon} \right) du^k \left(\frac{\partial}{\partial u^i_\epsilon} \right) \\ &= Y^k_\epsilon - \epsilon \left(X^j_\epsilon \frac{\partial Y^i_\epsilon}{\partial u^j_\epsilon} - Y^j_\epsilon \frac{\partial X^i_\epsilon}{\partial u^j_\epsilon} \right) \delta^k{}_i \\ &= Y^k_\epsilon - \epsilon \left(X^j_\epsilon \frac{\partial Y^k_\epsilon}{\partial u^j_\epsilon} - Y^j_\epsilon \frac{\partial X^k_\epsilon}{\partial u^j_\epsilon} \right) \\ &= Y^k_\epsilon - \epsilon \lambda^k_\epsilon \end{aligned} \tag{7.25}$$

Therefore,

$$\omega_{\sigma_{X,\epsilon}}((\sigma_{X,\epsilon})_* w) = \omega_k(u_\epsilon) Y^k(u_\epsilon) - \epsilon \omega_k(u_\epsilon) \lambda^k(u_\epsilon)$$

Now, we expand all terms in first order ϵ:

$$\begin{aligned} \omega_{\sigma_{X,\epsilon}}((\sigma_{X,\epsilon})_* Y) &= \omega_k(u_0 + \epsilon X_0) Y^k(u_0 + \epsilon X_0) \\ &\quad - \epsilon \omega_k(u_0 + \epsilon X_0) \lambda^k(u_0 + \epsilon X_0) \\ &= \left(\omega_k(u_0) + \epsilon X^j_0 \frac{\partial \omega_k}{\partial u^j_0} \right) \left(Y^k(u_0) + \epsilon X^j_0 \frac{\partial Y^k}{\partial u^j_0} \right) \\ &\quad - \epsilon \left(\omega_k(u_0) + \epsilon X^j_0 \frac{\partial \omega_k}{\partial u^j_0} \right) \left(\lambda^k(u_0) + \epsilon X^j_0 \frac{\partial \lambda^k}{\partial u^j_0} \right) \\ &= \omega_0(Y_0) + \epsilon Y^k_0 X^j_0 \frac{\partial \omega_k}{\partial u^j_0} + \epsilon \omega_k(u_0) X^j_0 \frac{\partial Y^k_0}{\partial u^j_0} \\ &\quad - \epsilon \omega_k(u_0) \lambda^k(u_0) + O(\epsilon^2) \\ &= \omega_0(Y_0) + \epsilon Y^k_0 X^j_0 \frac{\partial \omega_k}{\partial u^j_0} + \epsilon \omega_k(u_0) Y^j_0 \frac{\partial X^k}{\partial u^j_0} + O(\epsilon^2) \end{aligned}$$

Change the indices and factor out the arbitrary vector Y:

$$\omega_{\sigma_{X,\epsilon}}((\sigma_{X,\epsilon})_* Y) - \omega_0(Y_0) = \epsilon Y_0^k \left(X_0^j \frac{\partial \omega_k}{\partial u_0^j} + \omega_j(u_0) \frac{\partial X^j}{\partial u_0^k} \right) + O(\epsilon^2)$$

Finally, using Eq. (7.23),

$$\mathcal{L}_X \omega = \left(X^j \frac{\partial \omega_k}{\partial u^j} + \omega_j \frac{\partial X^j}{\partial u^k} \right) du^k \tag{7.26}$$

7.3.2 Lie derivative of covectors using Leibniz rule

Consider a covector ω operating on a tangent vector Y, construct the function $\omega(Y)$ and then apply the Leibniz rule:

$$\mathcal{L}_X(\omega(Y)) = (\mathcal{L}_X \omega)(Y) + \omega(\mathcal{L}_X Y)$$

$$\begin{aligned}
\mathcal{L}_X(\omega(Y)) &= X^i \frac{\partial \omega_j Y^j}{\partial u^i} \\
&= X^i \omega_j \frac{\partial Y^j}{\partial u^i} + X^i Y^j \frac{\partial \omega_j}{\partial u^i} \\
&= (\mathcal{L}_X \omega)(Y) + \omega_j \left(X^i \frac{\partial Y^j}{\partial u^i} - Y^i \frac{\partial X^j}{\partial u^i} \right)
\end{aligned}$$

Making $(\mathcal{L}_X \omega)(Y)$ the subject,

$$(\mathcal{L}_X \omega)(Y) = X^i Y^j \frac{\partial \omega_j}{\partial u^i} + \omega_j Y^i \frac{\partial X^j}{\partial u^i}$$

Swapping the indices and factoring out Y^j,

$$(\mathcal{L}_X \omega)(Y) = \left(X^i \frac{\partial \omega_j}{\partial u^i} + \omega_i \frac{\partial X^i}{\partial u^j} \right) Y^j$$

With $Y^j = du^j(Y)$ and for arbitrary Y,

$$\mathcal{L}_X \omega = \left(X^i \frac{\partial \omega_j}{\partial u^i} + \omega_i \frac{\partial X^i}{\partial u^j} \right) du^j \tag{7.27}$$

7.4 Lie Derivative of the Metric Tensor and Killing Fields

The metric tensor plays a central role in differential geometry. Indeed, we can derive a vector field from the Lie derivative of the metric tensor. This vector field is called the Killing field, named after Wilhelm Karl Joseph Killing. The Killing field is derived by solving the equation for the vector field X in which

$$\mathcal{L}_X g = 0$$

The Killing field is, in general, not unique. We will discuss more about the Killing fields after we derive the Lie derivative of the metric tensor.

7.4.1 Metric tensor as tensor product

We write the metric tensor as a tensor product:

$$g = g_{st} du^s du^t$$

When operating on basis vectors, we obtain the components of the metric tensor in this basis:

$$\begin{aligned} g(\partial_{u^i}, \partial_{u^j}) &= g_{st} du^s du^t (\partial_{u^i}, \partial_{u^j}) \\ &= g_{st} du^s (\partial_{u^i}) du^t (\partial_{u^j}) \\ &= g_{st} \delta^s{}_i \delta^t{}_j \\ &= g_{ij} \end{aligned}$$

7.4.2 Lie derivative of the metric tensor

To derive the Lie derivative of the metric tensor, we use Eq. (7.23). First, we derive the pullback of the metric tensor with respect to the flow of X. Apply the metric tensor on the two vectors Y, Z:

$$((\sigma_{X,\epsilon})^* g_{\sigma_{X,\epsilon}(u_0)})(Y, Z) = g_{u_\epsilon}((\sigma_{X,\epsilon})_* Y, (\sigma_{X,\epsilon})_* Z)$$

Expanding the metric tensor in terms of a tensor product,

$$((\sigma_{X,\epsilon})^* g_{\sigma_{X,\epsilon}(u_0)})(Y, Z) = g_{st} du^s((\sigma_{X,\epsilon})_* Y) du^t((\sigma_{X,\epsilon})_* Z))$$

Using Eq. (7.25),

$$du^s((\sigma_{X,\epsilon})_* Y) = Y^s_\epsilon - \epsilon \left(X^j_\epsilon \frac{\partial Y^s_\epsilon}{\partial u^j_\epsilon} - Y^j_\epsilon \frac{\partial X^s_\epsilon}{\partial u^j_\epsilon} \right)$$

$$du^t((\sigma_{X,\epsilon})_* Z) = Z^t_\epsilon - \epsilon\left(X^j_\epsilon \frac{\partial Z^t_\epsilon}{\partial u^j_\epsilon} - Z^j_\epsilon \frac{\partial X^t_\epsilon}{\partial u^j_\epsilon}\right)$$

Expanding in first order ϵ, we can see that the second term of the above equations reduces to simply evaluating it at u_0:

$$\begin{aligned} du^s((\sigma_{X,\epsilon})_* Y) &= Y^s_0 + \epsilon X^j_0 \frac{\partial Y^s_0}{\partial u^j_0} - \epsilon\left(X^j_0 \frac{\partial Y^s_0}{\partial u^j_0} - Y^j_0 \frac{\partial X^s_0}{\partial u^j_0}\right) \\ &= Y^s_0 + \epsilon Y^j_0 \frac{\partial X^s_0}{\partial u^j_0} \end{aligned}$$

$$\begin{aligned} du^t((\sigma_{X,\epsilon})_* Z) &= Z^t_0 + \epsilon X^j_0 \frac{\partial Z^t_0}{\partial u^j_0} - \epsilon\left(X^j_0 \frac{\partial Z^t_0}{\partial u^j_0} - Z^j_0 \frac{\partial X^t_0}{\partial u^j_0}\right) \\ &= Z^t_0 + \epsilon Z^j_0 \frac{\partial X^t_0}{\partial u^j_0} \end{aligned}$$

Expand the $g_{st}(u_\epsilon)$ term to first order:

$$g_{st}(u_\epsilon) \approx g_{st}(u_0) + \epsilon X^k \frac{\partial g_{st}}{\partial u^k}$$

Assembling the tensor components and keeping the first-order ϵ terms,

$$\begin{aligned} &\lim_{\epsilon\to 0}\left(\frac{((\sigma_{X,\epsilon})^* g_{\sigma_{X,\epsilon}(u_0)}) - g_{u_0}}{\epsilon}\right)(Y, Z) \\ &= \left(Y^s Z^t \frac{\partial g_{st}}{\partial u^j_0} + g_{st} Z^t_0 Y^j_0 \frac{\partial X^s_0}{\partial u^j_0} + g_{st} Y^s_0 Z^j_0 \frac{\partial X^t_0}{\partial u^j_0}\right) \end{aligned}$$

Reindexing so that we can factor out Y and Z terms and dropping the subscripts for clarity,

$$\mathcal{L}_X g(Y, Z) = \left(X^k \frac{\partial g_{st}}{\partial u^k} + g_{kt} \frac{\partial X^k}{\partial u^s} + g_{sk} \frac{\partial X^k}{\partial u^t}\right) Y^s Z^t$$

Since Y and Z are arbitrary,

$$\mathcal{L}_X g = \left(X^k \frac{\partial g_{st}}{\partial u^k} + g_{kt} \frac{\partial X^k}{\partial u^s} + g_{sk} \frac{\partial X^k}{\partial u^t}\right) du^s du^t$$

7.4.3 Lie derivative of the metric tensor using Leibniz rule

Using the expression of the Lie derivative on the covector and using the Leibniz rule, we can derive the Lie derivative for the 2-tensor $g = g_{st} du^s du^t$.

From Eq. (7.26),

$$\mathcal{L}_X du^j = \frac{\partial X^j}{\partial u^k} du^k$$

$$\begin{aligned}
\mathcal{L}_X g &= \mathcal{L}_X (g_{st} du^s du^t) \\
&= (\mathcal{L}_X g_{st}) du^s du^t + g_{st} (\mathcal{L}_X du^s) du^t + g_{st} du^s (\mathcal{L}_X du^t) \\
&= \left(X^k \frac{\partial g_{st}}{\partial u^k} du^s du^t + g_{st} \frac{\partial X^s}{\partial u^k} du^k du^t + g_{st} \frac{\partial X^t}{\partial u^k} du^s du^k \right)
\end{aligned}$$

Swapping indices to make the differentials have the same indices,

$$\mathcal{L}_X g = \left(X^k \frac{\partial g_{st}}{\partial u^k} + g_{kt} \frac{\partial X^k}{\partial u^s} + g_{sk} \frac{\partial X^k}{\partial u^t} \right) du^s du^t$$

7.4.4 Killing fields

As described above, if $\mathcal{L}_X g = 0$ for some X, then X is the Killing field. Note that the Killing field is not unique. Using the expression for the Lie derivative of the metric tensor, we get the differential equation for the Killing field:

$$X^k \frac{\partial g_{st}}{\partial u^k} + g_{kt} \frac{\partial X^k}{\partial u^s} + g_{sk} \frac{\partial X^k}{\partial u^t} = 0$$

The following are the properties of Killing fields:

1. A linear combination of Killing fields is a Killing field.
2. A Lie bracket of Killing fields is a Killing field.
3. Given a Killing field X and a geodesic γ with velocity vector $\dot{\gamma}$, the inner product $X^i g_{ij} \dot{\gamma}^j$ is conserved along γ.

7.5 The Lie Bracket

The previous section worked out the Lie derivative from first principles. The expression of the Lie derivative can be obtained by using the commutator. A commutator is defined abstractly on two operators as

$$[X, Y] = XY - YX$$

To use the commutator on the tangent vectors, we need consecutive directional differentials:

$$X^i \frac{\partial}{\partial u^i} \left(Y^j \frac{\partial f}{\partial u^j} \right) = X^i \frac{\partial Y^j}{\partial u^i} \frac{\partial f}{\partial u^j} + X^i Y^j \frac{\partial^2 f}{\partial u^j \partial u^i} \tag{7.28}$$

Writing out explicitly,

$$\begin{aligned} XY &= X^i \frac{\partial Y^j}{\partial u^i}\frac{\partial}{\partial u^j} + X^i Y^j \frac{\partial^2}{\partial u^j \partial u^i} \\ YX_1 &= Y^j \frac{\partial X^i}{\partial u^j}\frac{\partial}{\partial u^i} + X^i Y^j \frac{\partial^2}{\partial u^j \partial u^i} \\ [X,Y] &= X^i \frac{\partial Y^j}{\partial u^i}\frac{\partial}{\partial u^j} - Y^j \frac{\partial X^i}{\partial u^j}\frac{\partial}{\partial u^i} \end{aligned} \tag{7.29}$$

By changing the dummy indices,

$$\begin{aligned} [X,Y] &= X^i \frac{\partial Y^j}{\partial u^i}\frac{\partial}{\partial u^j} - Y^i \frac{\partial X^j}{\partial u^i}\frac{\partial}{\partial u^j} \\ &= \left(X^i \frac{\partial Y^j}{\partial u^i} - Y^i \frac{\partial X^j}{\partial u^i} \right) \frac{\partial}{\partial u^j} \end{aligned} \tag{7.30}$$

Hence, the Lie derivative and Lie bracket really refer to the same thing:

$$\mathcal{L}_X Y = [X,Y] = D_X Y - D_Y X$$

In the last term, we use the torsion-free property of the covariant derivative.

7.5.1 Properties of the Lie bracket and Lie derivative

Lemma 7.5.1 (Properties of Lie derivatives) *Some properties of Lie derivatives and the Lie bracket are stated as follows:*

1. *Antisymmetric:*

$$\begin{aligned} \mathcal{L}_X Y &= -\mathcal{L}_Y X \\ [X,Y] &= -[X,Y] \end{aligned} \tag{7.31}$$

2. *Bilinear:*

$$\begin{aligned} [X, Y+Z] &= [X,Y] + [X,Z] \\ [X+Y, Z] &= [X,Z] + [Y,Z] \end{aligned}$$

3. *The Jacobi identity, i.e. the sum of cyclic permutations is zero:*

$$[X,[Y,Z]] + [Z,[X,Y]] + [Y,[Z,X]] = 0 \tag{7.32}$$

4. *Product rule:*

$$[X, fY] = X(f)Y + f[X,Y] \tag{7.33}$$

5. *Commutator of Lie derivatives:*

$$[\mathcal{L}_X, \mathcal{L}_Y] = \mathcal{L}_{[X,Y]}$$

Lie derivative can be written in covariant derivatives

Lemma 7.5.2 *The Lie derivative can be written in the form of covariant derivatives:*

$$\mathcal{L}_X Y = D_X Y - D_Y X = [X, Y] \qquad \text{(torsion-free property)}$$

The proof is as follows:

$$\begin{aligned}
\mathcal{L}_X Y &= \left(X^j \frac{\partial Y^i}{\partial u^j} - Y^j \frac{\partial X^i}{\partial u^j} \right) \frac{\partial}{\partial u^i} \\
&= \left(X^j (D_{\hat{e}_j} Y)^i - \Gamma^i{}_{sj} X^j Y^s - Y^j (D_{\hat{e}_j} X)^i + \Gamma^i{}_{sj} X^j Y^s \right) \frac{\partial}{\partial u^i} \\
&= \left(X^j (D_{\hat{e}_j} Y)^i - Y^j (D_{\hat{e}_j} X)^i \right) \partial_{u^i} \\
&= D_X Y - D_Y X = [X, Y] \qquad \text{(torsion free property)}
\end{aligned}$$

Lie derivative on scalars

$$\mathcal{L}_X f = X^i \frac{\partial f}{\partial u^i} = X(f)$$

7.5.2 Commuting flows and the Lie bracket

Definition 7.5.1 Let X and Y be two vector fields in U. Let their respective flows be $\sigma_{X,t}$ and $\sigma_{Y,s}$. The flows commute if for all $u \in U$,

$$\sigma_{X,t} \circ \sigma_{Y,s}(u) = \sigma_{Y,s} \circ \sigma_{X,t}(u)$$

Theorem 7.5.1 (commuting flows) *Let X and Y be two vector fields in U with the respective flows as $\sigma_{X,t}$ and $\sigma_{Y,s}$. The flows commute iff $[X, Y] = 0$.*

A proof of the commuting flows theorem is as follows. First, we recognize that if infinitesimal flows are commuting, we can assemble infinitesimal flows into finite flows, and the finite flows still commute. That is, finite flows are made up of many infinitesimal flows. Hence, it is sufficient to consider infinitesimal flows. Figure 7.6 shows two paths ABD and ACD'. We prove both ways: first, assume that the two paths meet, that is, $u_D = u_{D'}$. Then, show that $[X, Y] = 0$, and then we show the converse.

We derive the locations of u_D and $u_{D'}$ by applying the Taylor series expansion to first-order infinitesimal flows along X by $t \ll 1$ and Y by $s \ll 1$.

$$\begin{aligned}
\sigma^i_{X,t}(u) &= u^i + \frac{d\sigma^i_{X,t}}{dt} t + O(t^2) \\
&= u^i + X^i t + O(t^2) \qquad (7.34)
\end{aligned}$$

Figure 7.6: Given two vector fields X and Y and starting point A, flow by $\sigma_{y,t}(A)$ to C and then flow by $\sigma_{X,t}(C)$ to D'. Alternatively we can flow by $\sigma_{X,t}(A)$ to B and then $\sigma_{Y,t}(B)$ to B and then $\sigma_{Y,t}(B)$ to D. $D = D'$ if and only if $[X, Y] = 0$.

Similarly, for flows in the Y direction,

$$\sigma^i_{Y,s}(u) \quad = \quad u^i + Y^i s + O(s^2) \tag{7.35}$$

Along the path ABD,

$$u^i_B = u^i_A + X^i t + O(t^2)$$

$$\begin{aligned} u^i_D &= u^i_B + Y^i(u_B)s + O(t^2) \\ &= u^i_A + X^i t + Y^i(u_A + Xt)s + O(t^2) \\ &\approx u^i_A + X^i t + Y^i(u_A)s + \left.\frac{\partial Y^i}{\partial u^j}\right|_{u_A} X^j \partial_{u^i} ts \\ u^i_D &\approx u^i_A + X^i t + Y^i s + \left.\frac{\partial Y^i}{\partial u^j}\right|_{u_A} X^j \partial_{u^i} ts \end{aligned}$$

Using a similar procedure, we get

$$u^i_{D'} \approx u^i_A + X^i t + Y^i s + \left.\frac{\partial X^i}{\partial u^j}\right|_{u_A} Y^j \partial_{u^i} ts$$

Now, we are ready to prove that if $u_D = u_{D'}$, then $[X, Y] = 0$. Equate u_D and $u_{D'}$. Then, to the first-order approximation in t and s,

$$\begin{aligned} u^i_A + X^i t + Y^i s + \frac{\partial Y^i}{\partial u^j} X^j \partial_{u^i} ts &= u^i_A + X^i t + Y^i s + \frac{\partial X^i}{\partial u^j} Y^j \partial_{u^i} ts \\ \frac{\partial Y^i}{\partial u^j} X^j &= \frac{\partial X^i}{\partial u^j} Y^j \\ [X, Y] &= 0 \end{aligned}$$

If we start with $[X, Y] = 0$, then

$$u^i_D - u^i_{D'} \approx \left(\frac{\partial Y^i}{\partial u^j} X^j - \frac{\partial X^i}{\partial u^j} Y^j\right) ts \partial_{u^i} = [X, Y]^i ts = 0$$

Exercise 7.5.1

1. Prove the properties of Lie derivatives and the Lie bracket as given by

   Lemma 7.5.1 (Properties of Lie derivatives) *Some properties of Lie derivatives and the Lie bracket are stated as follows:*

   1. Antisymmetric:

   $$\mathcal{L}_X Y = -\mathcal{L}_Y X$$
   $$[X, Y] = -[X, Y] \tag{7.31}$$

   2. Bilinear:

   $$\begin{aligned} [X, Y + Z] &= [X, Y] + [X, Z] \\ [X + Y, Z] &= [X, Z] + [Y, Z] \end{aligned}$$

   3. The Jacobi identity, i.e. the sum of cyclic permutations is zero:

   $$[X, [Y, Z]] + [Z, [X, Y]] + [Y, [Z, X]] = 0 \tag{7.32}$$

   4. Product rule:

   $$[X, fY] = X(f)Y + f[X, Y] \tag{7.33}$$

   5. Commutator of Lie derivatives:

   $$[\mathcal{L}_X, \mathcal{L}_Y] = \mathcal{L}_{[X,Y]}$$

   Solution: To show $[X, Y] = -[Y, X]$,

   $$[X, Y] = XY - YX = -(YX - XY) = -[Y, X]$$

   To show bilinearity, since X and Y are differential operators, they are linear in their inputs, that is,

   $$X(Y + Z) = XY + XZ$$

   Similarly,

   $$(X + Y)(Z) = XZ + YZ$$

Putting these results together,

$$\begin{aligned}[X, Y+Z] &= X(Y+Z) - (Y+Z)X = XY + XZ - YX - ZX \\ &= [X, Y] + [X, Z]\end{aligned}$$

Also,

$$[X+Y, Z] = -[Z, X+Y] = -[Z, X] - [Z, Y] = [X, Z] + [Y, Z]$$

To derive the Jacobi identity, just expand out the commutator:

$$\begin{aligned}&[X,[Y,Z]] + [Z,[X,Y]] + [Y,[Z,X]] \\ &= X[Y,Z] - [Y,Z]X + Z[X,Y] - [X,Y]Z + Y[Z,X] - [Z,X]Y \\ &= XYZ - XZY - YZX + ZYX + ZXY - ZYX - XYZ \\ &\quad + YXZ + YZX - YXZ - ZXY + XZY \\ &= 0\end{aligned}$$

For the product rule, $[X, fY] = X(f)Y + f[X,Y]$, expand out the commutator, and using the product rule for differentiation $X(fY) = X(f)Y + fXY$,

$$\begin{aligned}[X, fY] &= X(fY) - fYX \\ &= X(f)Y + fXY - fYX \\ &= X(f)Y + f[X,Y]\end{aligned}$$

To show the commutator of Lie derivatives, there are three main steps. First, assign an arbitrary Z for the commutator to act on. Second, use bilinear properties. Lastly, use the Jacobi identity to simplify:

$$\begin{aligned}[\mathcal{L}_X, \mathcal{L}_Y]Z &= \mathcal{L}_X\mathcal{L}_Y Z - \mathcal{L}_Y\mathcal{L}_X Z \\ &= \mathcal{L}_X[Y,Z] - \mathcal{L}_Y[X,Z] \\ &= [X,[Y,Z]] - [Y,[X,Z]] \\ &= [X,[Y,Z]] + [Y,[Z,X]] \text{ (bilinear)}\end{aligned}$$

Using the Jacobi identity,

$$[Y,[Z,X]] = -[X,[Y,Z]] - [Z,[X,Y]]$$

and substituting into the above equation, we see that one term cancels:

$$\begin{aligned}[\mathcal{L}_X, \mathcal{L}_Y]Z &= [X,[Y,Z]] - [X,[Y,Z]] - [Z,[X,Y]] \\ &= -[Z,[X,Y]] \\ &= [[X,Y],Z] \\ &= \mathcal{L}_{[X,Y]}Z\end{aligned}$$

Since Z is arbitrary, we get

$$[\mathcal{L}_X, \mathcal{L}_Y] = \mathcal{L}_{[X,Y]}$$

7.6 Summary

Definition 7.6.1 (Flow path) Given a vector field X and a point $x_0 \in M$, with coordinate representation $u) \in U$, a flow path $\gamma(t)$ is given by a path traced out by a parameter t such that

$$\begin{aligned} \gamma(0) &= x(u_0) \\ \dot{\gamma}(t) &= X(\gamma(t)) \end{aligned}$$

Denote $\sigma_{X,t}(x_0) = \gamma(t)$.

Definition 7.6.2 (Pullback) Given two manifolds U and $\hat{U}$, a mapping $\varphi : M \mapsto \hat{M}$ and a function $\hat{f} : \hat{M} \mapsto \mathbb{R}$, the pullback is defined as

$$f = \hat{f} \circ \varphi = \varphi^* \hat{f}$$

Definition 7.6.3 (Pushforward) Given two manifolds U and $\hat{U}$, a mapping $\varphi : U \mapsto \hat{M}$ and a vector $X \in T_u U$, the pushforward of X under φ is such that the pushforward vector $\hat{X}$ satisfies

$$X(\hat{f} \circ \varphi) = \hat{X}(\hat{f})$$

for any $\hat{f} \in C^\infty$. We denote

$$\hat{X} = \varphi_* X$$

Note that the mapping $\varphi_* : T_u U \mapsto T_{\varphi(u)} \hat{U}$.

Theorem 7.1.1 *Let X be a smooth vector field on U. Then, the flow path generated by X starting from u_0 and the pushforward of X under $\sigma_{X,t}$ satisfy,*[3]

$$\sigma_{X,t}(u_0)_* X_{u_0} = X_{\sigma_{X,t}(u_0)} \tag{7.3}$$

[3] See John Lee, *Introduction to Smooth Manifolds*, p. 442 [8].

Theorem 7.1.2 *Let X be a vector field in U with flow $\sigma_{X,t}$ and $\hat{X}$ be a vector field in $\hat{U}$ with flow $\sigma_{\hat{X},t}$. Let $\varphi : U \mapsto \hat{U}$. Then, φ_* relates to the flows by*[4]

$$\sigma_{\hat{X},t} \circ \varphi = \varphi \circ \sigma_{X,t} \iff \varphi_* X = \hat{X}. \tag{7.4}$$

Definition 7.6.4 (Lie derivative) The Lie derivative of a vector field Y in the direction of another vector field is X,

$$\mathcal{L}_X Y = \left(X^j \frac{\partial Y^i}{\partial u^j} - Y^j \frac{\partial X^i}{\partial u^j} \right) \partial_{u^i}$$

The Lie derivative of a covector field ω is

$$\mathcal{L}_X \omega = \left(X^i \frac{\partial \omega_j}{\partial u^i} + \omega_i \frac{\partial X^i}{\partial u^j} \right) du^j$$

Lemma 7.6.1 *The Lie derivative is equal to the Lie bracket:*

$$\mathcal{L}_X Y = [X, Y]$$

Lemma 7.5.1 (Properties of Lie derivatives) *Some properties of Lie derivatives and the Lie bracket are stated as follows:*

1. *Antisymmetric:*

$$\begin{aligned} \mathcal{L}_X Y &= -\mathcal{L}_Y X \\ [X, Y] &= -[X, Y] \end{aligned} \tag{7.31}$$

2. *Bilinear:*

$$\begin{aligned} [X, Y + Z] &= [X, Y] + [X, Z] \\ [X + Y, Z] &= [X, Z] + [Y, Z] \end{aligned}$$

3. *The Jacobi identity, i.e. the sum of cyclic permutations is zero:*

$$[X, [Y, Z]] + [Z, [X, Y]] + [Y, [Z, X]] = 0 \tag{7.32}$$

4. *Product rule:*

$$[X, fY] = X(f)Y + f[X, Y] \tag{7.33}$$

5. *Commutator of Lie derivatives:*

$$[\mathcal{L}_X, \mathcal{L}_Y] = \mathcal{L}_{[X,Y]}$$

[4] See John Lee, *Introduction to Smooth Manifolds*, p. 468 [8].

Lemma 7.5.2 *The Lie derivative can be written in the form of covariant derivatives:*

$$\mathcal{L}_X Y = D_X Y - D_Y X = [X, Y] \qquad \text{(torsion-free property)}$$

Definition 7.5.1 Let X and Y be two vector fields in U. Let their respective flows be $\sigma_{X,t}$ and $\sigma_{Y,s}$. The flows commute if for all $u \in U$,

$$\sigma_{X,t} \circ \sigma_{Y,s}(u) = \sigma_{Y,s} \circ \sigma_{X,t}(u)$$

Theorem 7.5.1 (commuting flows) *Let X and Y be two vector fields in U with the respective flows as $\sigma_{X,t}$ and $\sigma_{Y,s}$. The flows commute iff $[X, Y] = 0$.*

Chapter 8

Riemann Curvature Tensor

In this chapter, we explore the curvature of a manifold. How do we know if a manifold is "curved"? To answer this question, we shall make use of the foundations built in the previous chapters. First, we lay down the first principles of measuring if a manifold is curved. We say that the manifold is curved if, upon parallel transport of a vector along an infinitesimally small closed path, the transported vector and the original vector are not equal. The notion of infinitesimally small suggests that we need to perform differentiation. Recall that one calculates the curvature of a function by taking its second derivative.

8.1 Riemann Curvature Tensor Using Parallel Transport

We shall make precise the measure of curvature at a point $u \in U$ by parallel transport along two basis tangent vectors. Given a manifold U with dimension d_U, a closed path can be formed by arbitrarily choosing two basis tangent vectors and forming a "curved" parallelogram. Figure 8.2 illustrates a closed parallelogram using arbitrarily chosen ∂_{u^1} and ∂_{u^2}. Note that the paths are not exactly straight lines if the manifold is curved. There are two paths from point A to point D: one through γ_{ABD} and another through γ_{ACD}. If we arbitrarily transport a vector Z_A via these two paths, we end up with two vectors: Z_{ABD} and Z_{ACD}. We define the curvature at point A as the difference between Z_{ABD} and Z_{ACD} divided by the area enclosed

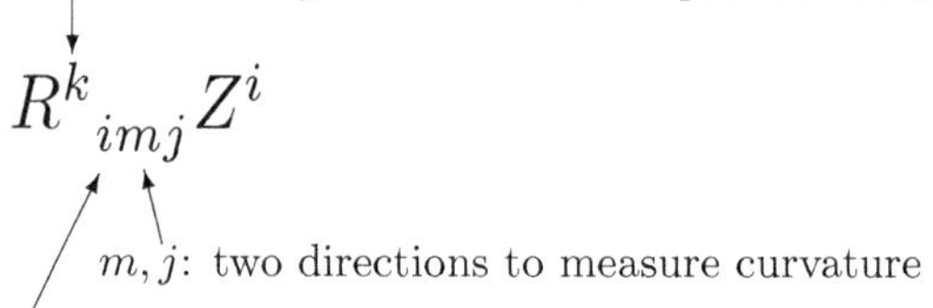

Figure 8.1: Explanation of the components of the Riemann curvature tensor.

by the path Δ. Take the k components of Z_{ABD} and Z_{ACD}:

$$R^k{}_{i12} Z^i = \lim_{\Delta \to 0} \frac{Z^k_{ABD} - Z^k_{ACD}}{\Delta} \tag{8.1}$$

This is called the Riemann curvature tensor. Now, we explain Eq. (8.1) to clarify our understanding of the physics underlying this equation. First, the RHS of this equation depends on: (1) the paths ∂_{u^1} and ∂_{u^2}, (2) the component of the transported vectors k, and (3) the vector that we started with at point A. Obviously, different paths give different values for the change of vectors. If we start off with a different Z, then the difference in parallel transport varies. However, the curvature at a point should be a quantity that depends on the geometry at the point itself. It should not depend on what vector we use for parallel transport. To make the LHS and RHS of Eq. (8.1) consistent, we "invent" a quantity that is independent of the starting vector and contract it with the starting vector: $R^k{}_{i12} Z^i$. This quantity must depend on the paths ∂_{u^1} and ∂_{u^2} and the component k of Z_{ABD} and Z_{ACD}. We write this quantity as $R^k{}_{i12}$. Figure 8.1 explains the role of each indices in R^k_{imj}. For the arbitrary paths ∂_{u^m} and ∂_{u^j}, we have $R^k{}_{imj}$. Now, we have extracted all path dependencies, components, and original vector dependencies. $R^k{}_{imj}$ is now an object that specifies the curvature properties of the manifold at a point. We can consider this as a component of the Riemann curvature tensor. Its value changes under coordinate transformations. k is the component of the change of vectors transported in the direction of ∂_{u^m} and ∂_{u^j}. i is the index of contraction with the original vector.

We can also consider the Riemann curvature tensor as a tensor that takes in two directional vectors, and we apply it to one starting vector to give a resultant vector. Writing R in covector and vector basis,

$$R = R^k{}_{imj} du^m du^j du^i \partial_{u^k}$$

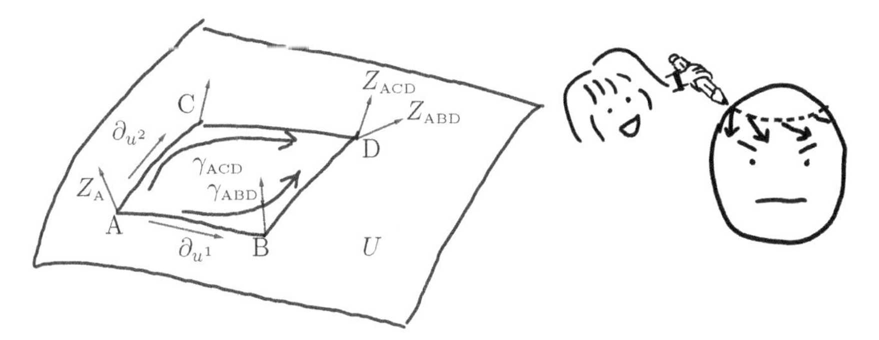

Figure 8.2: Deriving the Riemann curvature tensor using parallel transport.

such that the Riemann curvature tensor operates on three tangent vectors to output a tangent vector, $R(X,Y)Z \in T_u U$.

8.1.1 Parallel transport equations

Recall that the parallel transport equation along a path is (Eq. (5.31))

$$\frac{D_{\dot{\gamma}} Z(t)}{dt} = 0 \qquad \forall t \in [0,1]$$

In the above, the covariant derivative on a path is given in Eq. (5.10) as

$$\frac{D_{\dot{\gamma}} Z}{dt} = \left(D_{e^u_j} Z\right) \dot{\gamma}^j = 0$$

The covariant derivative is given by Eq. (5.3):

$$D_{\partial_{u^j}} Z = \left(\frac{\partial Z^k}{\partial u^j} + Z^i \Gamma^k{}_{ij}\right) \partial_{u^k} = 0$$

Since the basis tangent vectors are independent, each component of the covariant derivative equals zero:

$$\frac{\partial Z^k}{\partial u^j} = -Z^i \Gamma^k{}_{ij} \tag{8.2}$$

8.1.2 Path definitions and area

Using Eq. (8.2) as the key equation, we can integrate to get the components of Z along γ_{ABD} and γ_{ACD} and then calculate the differences between Z_{ABD} and Z_{ACD}. For the integration, we need to define infinitesimal paths using coordinate representation. Let the coordinates of points A, B, and C be u_A, u_B, and u_C, respectively:

$$\begin{aligned}
u^i_{AB}(t) &= u_A + \epsilon_1 t \partial_{u^1} \\
u^i_{BD}(t) &= u_A + \epsilon_1 \partial_{u^1} + \epsilon_2 t \partial_{u^2} \\
u^i_{AC}(t) &= u_A + \epsilon_2 t \partial_{u^2} \\
u^i_{CD}(t) &= u_A + \epsilon_2 \partial_{u^2} + \epsilon_1 t \partial_{u^1}
\end{aligned}$$

We take the limits $\epsilon_1 \to 0$ and $\epsilon_2 \to 0$. In these limits, $\Delta = \epsilon_1 \epsilon_2$ is the infinitesimal area enclosed by the paths. Note that the two end points meet, i.e. $u^i_{BD}(1) = u^i_{CD}(1)$.

8.1.3 Integration of component along paths

At this point, we note that two tangent vectors cannot be compared at different points on a manifold. However, we can compare their components because components are simply scalar functions on the manifold. We can also perform integration of the components of vectors (which are scalar functions) along a path on the manifold. Integrate along γ_{ABD},

$$Z^k_{ABD} - Z^k_A = -\int_A^B Z^i \Gamma^k{}_{i1} du_{AB}(t) - \int_B^D Z^i \Gamma^k{}_{i2} du_{BD}(t) \tag{8.3}$$

Recall that Eq. (3.19) expresses arc length as

$$du = \sqrt{\dot{u}(t) \cdot \dot{u}(t)} dt = |\dot{u}(t)| dt$$

where u along different paths are linear in t, e.g. $|\dot{u}^i_{AB}(t)| = \epsilon_1$, $|\dot{u}^i_{BD}(t)| = \epsilon_2$. Writing $\varpi^k_1 = Z^i \Gamma^k{}_{i1}$, the first term in Eq. (8.4) becomes

$$\begin{aligned}
\epsilon_1 \int_0^1 \varpi^k_1(u_A + \epsilon_1 t \partial_{u^1}) dt &= \epsilon_1 \int_0^1 \varpi^k_1(u_A) + \epsilon_1 t \frac{\partial \varpi^k_1}{\partial u^1} dt \\
&= \epsilon_1 \varpi^k_1(u_A) + \frac{\epsilon_1^2}{2} \frac{\partial \varpi^k_1}{\partial u^1}\bigg|_{u_A}
\end{aligned}$$

Using the same procedure, the second term in Eq. (8.4) becomes

$$\begin{aligned}
\epsilon_2 \int_0^1 \varpi^k_2(u_A + \epsilon_1 \partial_{u^1} + \epsilon_2 t \partial_{u^2}) dt &= \epsilon_2 \int_0^1 \varpi^k_2(u_A) + \epsilon_1 \frac{\partial \varpi^k_2}{\partial u^1} + \epsilon_2 t \frac{\partial \varpi^k_2}{\partial u^2} dt \\
&= \epsilon_2 \varpi^k_2(u_A) + \epsilon_1 \epsilon_2 \frac{\partial \varpi^k_2}{\partial u^1}\bigg|_{u_A} + \frac{\epsilon_2^2}{2} \frac{\partial \varpi^k_2}{\partial u^2}\bigg|_{u_A}
\end{aligned}$$

Putting together the two terms,

$$
\begin{aligned}
&Z^k_{ABD} - Z^k_A \qquad\qquad (8.4)\\
&= -\epsilon_1 \varpi^k_1(u_A) - \frac{\epsilon_1^2}{2}\frac{\partial \varpi^k_1}{\partial u^1}\bigg|_{u_A} - \epsilon_2 \varpi^k_2(u_A) - \epsilon_1\epsilon_2 \frac{\partial \varpi^k_2}{\partial u^1}\bigg|_{u_A} - \frac{\epsilon_2^2}{2}\frac{\partial \varpi^k_2}{\partial u^2}\bigg|_{u_A}
\end{aligned}
$$

Integrating along γ_{ACD},

$$
Z^k_{ACD} - Z^k_A = -\int_A^C Z^i \Gamma^k{}_{i2} d\gamma_{AC} - \int_C^D Z^i \Gamma^k{}_{i1} d\gamma_{CD}
$$

The first and second terms of this integral become, respectively,

$$
\begin{aligned}
\epsilon_2 \int_0^1 \varpi^k_2(u_A + \epsilon_2 t \partial_{u^2}) dt &= \epsilon_2 \int_0^1 \varpi^k_2(u_A) + \epsilon_2 t \frac{\partial \varpi^k_2}{\partial u^2} dt\\
&= \epsilon_2 \varpi^k_2(u_A) + \frac{\epsilon_2^2}{2}\frac{\partial \varpi^k_2}{\partial u^2}\bigg|_{u_A}
\end{aligned}
$$

$$
\begin{aligned}
&\epsilon_1 \int_0^1 \varpi^k_1(u_A + \epsilon_2 \partial_{u^2} + \epsilon_1 t \partial_{u^1}) dt\\
&= \epsilon_1 \int_0^1 \varpi^k_1(u_A) + \epsilon_2 \frac{\partial \varpi^k_1}{\partial u^2} + \epsilon_1 t \frac{\partial \varpi^k_1}{\partial u^1} dt\\
&= -\epsilon_1 \varpi^k_1(u_A) - \epsilon_1\epsilon_2 \frac{\partial \varpi^k_1}{\partial u^2}\bigg|_{u_A} + \frac{\epsilon_1^2}{2}\frac{\partial \varpi^k_1}{\partial u^1}\bigg|_{u_A}
\end{aligned}
$$

Along the path ACD, we have

$$
\begin{aligned}
&Z^k_{ACD} - Z^k_A \qquad\qquad (8.5)\\
&= -\epsilon_2 \varpi^k_2(u_A) - \frac{\epsilon_2^2}{2}\frac{\partial \varpi^k_2}{\partial u^2}\bigg|_{u_A} \epsilon_1 \varpi^k_1(u_A) - \epsilon_1\epsilon_2 \frac{\partial \varpi^k_1}{\partial u^2}\bigg|_{u_A} - \frac{\epsilon_1^2}{2}\frac{\partial \varpi^k_1}{\partial u^1}\bigg|_{u_A}
\end{aligned}
$$

Now that we have evaluated everything at u_A, we can drop the subscript used to indicate evaluation at u_A. Subtracting the integrals of both paths (Eqs. (8.4) and (8.5)), we find that all terms except the $\epsilon_1\epsilon_2$ term cancel:

$$
Z^k_{ABD} - Z^k_{ACD} = \epsilon_1\epsilon_2 \left(\frac{\partial \varpi^k_1}{\partial u^2} - \frac{\partial \varpi^k_2}{\partial u^1} \right)
$$

Using Eq. (8.1) and taking the limits $\epsilon_1 \to 0$ and $\epsilon_2 \to 0$,

$$
R^k{}_{i12} Z^i = \lim_{\epsilon_1, \epsilon_2 \to 0} \frac{Z^k_{ABD} - Z^k_{ACD}}{\epsilon_1 \epsilon_2} = \frac{\partial \varpi^k_1}{\partial u^2} - \frac{\partial \varpi^k_2}{\partial u^1}
$$

Recall that $\varpi_j^k = Z^i\Gamma^k{}_{ij}$, $j = 1, 2$ and use the product rule to differentiate:

$$\begin{aligned} R^k{}_{i12}Z^i &= \left(\frac{\partial Z^i}{\partial u^2}\Gamma^k{}_{i1} + Z^i\frac{\partial \Gamma^k{}_{i1}}{\partial u^2}\right) - \left(\frac{\partial Z^i}{\partial u^1}\Gamma^k{}_{i2} + Z^i\frac{\partial \Gamma^k{}_{i2}}{\partial u^1}\right) \\ &= \left(-Z^l\Gamma^i{}_{l2}\Gamma^k{}_{i1} + Z^i\frac{\partial \Gamma^k{}_{i1}}{\partial u^2}\right) - \left(-Z^l\Gamma^i{}_{l1}\Gamma^k{}_{i2} + Z^i\frac{\partial \Gamma^k{}_{i2}}{\partial u^1}\right) \end{aligned}$$

Swapping the dummy indices $i \leftrightarrow l$ to factorize out Z^k,

$$\begin{aligned} R^k{}_{i12}Z^i &= \left(-\Gamma^l{}_{i2}\Gamma^k{}_{l1} + \frac{\partial \Gamma^k{}_{i1}}{\partial u^2} + \Gamma^l{}_{i1}\Gamma^k{}_{l2} - \frac{\partial \Gamma^k{}_{i2}}{\partial u^1}\right) Z^i \\ &= \left(\Gamma^l{}_{i1}\Gamma^k{}_{l2} - \Gamma^l{}_{i2}\Gamma^k{}_{l1} + \frac{\partial \Gamma^k{}_{i1}}{\partial u^2} - \frac{\partial \Gamma^k{}_{i2}}{\partial u^1}\right) Z^i \end{aligned}$$

Then, since ∂_{u^1} and ∂_{u^2} are arbitrary, replace them with the indices j and m, respectively. As Z^k is also arbitrary, omit it to represent the curvature components, $R^k{}_{imj}$:

$$R^k{}_{imj} = \Gamma^k{}_{lm}\Gamma^l{}_{ij} - \Gamma^k{}_{lj}\Gamma^l{}_{im} + \frac{\partial \Gamma^k{}_{ij}}{\partial u^m} - \frac{\partial \Gamma^k{}_{im}}{\partial u^j} \tag{8.6}$$

Finally, if we want to use the Riemann curvature tensor as an operator that takes in tangent vectors, we write it as

$$R = R^k{}_{imj}du^m du^j du^i \partial_{u^k} \tag{8.7}$$

such that the Riemann curvature tensor operates on 3 tangent vectors to output a tangent vector, $R(X, Y)Z \in T_uU$.

8.2 Riemann Curvature Tensor Using Covariant Derivative

Another method to derive the Riemann curvature tensor is through the commutator of a covariant derivative in the direction of basis tangent vectors. This means that there is indeed a relationship between parallel transport and commutator of covariant derivatives. We discuss this relationship after the derivation. The commutator of a covariant derivative is given by

$$[D_{\partial_{u^m}}, D_{\partial_{u^j}}]Z = (D_{\partial_{u^m}}D_{\partial_{u^j}} - D_{\partial_{u^j}}D_{\partial_{u^m}})Z$$

Differentiating twice,

$$D_{\partial_{u^j}}Z = \left(\frac{\partial Z^k}{\partial u^j} + Z^i\Gamma^k{}_{ij}\right)\partial_{u^k}$$

$$D_{\partial_{u^m}} D_{\partial_{u^j}} Z = \left(\frac{\partial}{\partial u^m} \left(\frac{\partial Z^k}{\partial u^j} + Z^i \Gamma^k{}_{ij} \right) + \left(\frac{\partial Z^t}{\partial u^j} + Z^i \Gamma^t{}_{ij} \right) \Gamma^k{}_{tm} \right) \partial_{u^k}$$

$$= \left(\frac{\partial^2 Z^k}{\partial u^m \partial u^j} + \frac{\partial Z^i \Gamma^k{}_{ij}}{\partial u^m} + \frac{\partial Z^t}{\partial u^j} \Gamma^k{}_{tm} + Z^i \Gamma^t{}_{ij} \Gamma^k{}_{tm} \right) \partial_{u^k} \tag{8.8}$$

Evaluate $D_{\partial_{u^m}} D_{\partial_{u^j}}$ by swapping the indices $j \leftrightarrow m$:

$$D_{\partial_{u^j}} D_{\partial_{u^m}} Z = \left(\frac{\partial^2 Z^k}{\partial u^j \partial u^m} + \frac{\partial Z^i \Gamma^k{}_{im}}{\partial u^j} + \frac{\partial Z^t}{\partial u^m} \Gamma^k{}_{tj} + Z^i \Gamma^t{}_{im} \Gamma^k{}_{tj} \right) \partial_{u^k}$$

Swap the indices i and t in the second term and t and i in the third term:

$$D_{\partial_{u^j}} D_{\partial_{u^m}} Z = \left(\frac{\partial^2 Z^k}{\partial u^j \partial u^m} + \frac{\partial Z^t \Gamma^k{}_{tm}}{\partial u^j} + \frac{\partial Z^i}{\partial u^m} \Gamma^k{}_{ij} + Z^i \Gamma^t{}_{im} \Gamma^k{}_{tj} \right) \partial_{u^k} \tag{8.9}$$

Perform the commutation operation by subtracting Eq. (8.9) from Eq. (8.8):

$$[D_{\partial_{u^m}} D_{\partial_{u^j}} - D_{\partial_{u^j}} D_{\partial_{u^m}}] Z \left(\frac{\partial Z^i \Gamma^k{}_{ij}}{\partial u^m} - \frac{\partial Z^i}{\partial u^m} \Gamma^k{}_{ij} + \frac{\partial Z^t}{\partial u^j} \Gamma^k{}_{tm} \right.$$
$$\left. - \frac{\partial Z^t \Gamma^k{}_{tm}}{\partial u^j} + Z^i \Gamma^t{}_{ij} \Gamma^k{}_{tm} - Z^i \Gamma^t{}_{im} \Gamma^k{}_{tj} \right) \partial_{u^k}$$

The first and second terms are part of the product rule of differentiation. In particular, the first term can be reduced to $Z^s \partial \Gamma^k{}_{sj} / \partial u^i$, and the dummy index t can be changed to s in the second term:

$$[D_{\partial_{u^m}} D_{\partial_{u^j}} - D_{\partial_{u^j}} D_{\partial_{u^m}}] Z$$
$$= \left(\Gamma^t{}_{ij} \Gamma^k{}_{tm} - \Gamma^t{}_{im} \Gamma^k{}_{tj} + \frac{\partial \Gamma^k{}_{ij}}{\partial u^m} - \frac{\partial \Gamma^k{}_{im}}{\partial u^j} \right) Z^i \partial_{u^k}$$

By comparing with Eq. (8.6), we rearrange some terms to arrive at

$$\begin{aligned} R^k{}_{imj} &= [D_{\partial_{u^j}} D_{\partial_{u^m}} - D_{\partial_{u^m}} D_{\partial_{u^j}}]^k_i \\ &= \Gamma^k{}_{lm} \Gamma^l{}_{ij} - \Gamma^k{}_{lj} \Gamma^l{}_{im} + \frac{\partial \Gamma^k{}_{ij}}{\partial u^m} - \frac{\partial \Gamma^k{}_{im}}{\partial u^j} \end{aligned}$$

8.2.1 General form of the Riemann curvature tensor with covariant derivative

Using the properties of the covariant derivative,

$$\begin{aligned} D_X Z &= D_{X^i \partial_{u^i}} Z = X^i D_{\partial_{u^i}} Z \\ D_{fX} Z &= f D_X Z + (D_X f) Z \end{aligned}$$

$$\begin{aligned} D_X D_Y Z &= D_{X^i \partial_{u^i}} D_{Y^j \partial_{u^j}} Z \\ &= X^i D_{\partial_{u^i}} (Y^j D_{\partial_{u^j}} Z) \\ &= X^i Y^j D_{\partial_{u^i}} D_{\partial_{u^j}} Z + X^i (\partial_{u_i} Y^j) D_{\partial_{u^j}} Z \end{aligned}$$

$$\begin{aligned} D_Y D_X Z &= D_{Y^j \partial_{u^j}} D_{X^i \partial_{u^i}} Z \\ &= Y^j D_{\partial_{u^j}} (X^i D_{\partial_{u^i}} Z) \\ &= Y^j X^i D_{\partial_{u^j}} D_{\partial_{u^i}} Z + Y^j (\partial_{u_j} X^i) D_{\partial_{u^i}} Z \end{aligned}$$

Subtracting and swapping indices,

$$\begin{aligned} D_X D_Y - D_Y D_X &= Y^j X^i [D_{\partial_{u^i}}, D_{\partial_{u^j}}] + X^i (\partial_{u_i} Y^j) D_{\partial_{u^j}} - Y^i (\partial_{u_i} X^j) D_{\partial_{u^j}} \\ &= Y^j X^i [D_{\partial_{u^i}}, D_{\partial_{u^j}}] + D_{[X,Y]} \\ &= Y^j X^i R(\partial_{u^i}, \partial_{u^j}) + D_{[X,Y]} \\ X^i Y^j R(\partial_{u^i}, \partial_{u^j}) &= [D_X, D_Y] - D_{[X,Y]} \end{aligned}$$

When we write the Riemann curvature tensor as tensor products of covectors, we see that it is multilinear. Therefore,

$$X^i Y^j R(\partial_{u^i}, \partial_{u^j}) = R(X^i \partial_{u^i}, Y^j \partial_{u^j}) = R(X, Y)$$

Hence,

$$R(X, Y) = [D_X, D_Y] - D_{[X,Y]} \tag{8.10}$$

8.3 Properties of $R^k{}_{imj}$ and R_{kimj}

Given $R^k{}_{imj}$,

$$R^k{}_{imj} = \Gamma^k{}_{lm}\Gamma^l{}_{ij} - \Gamma^k{}_{lj}\Gamma^l{}_{im} + \frac{\partial \Gamma^k{}_{ij}}{\partial u^m} - \frac{\partial \Gamma^k{}_{im}}{\partial u^j} \tag{8.6}$$

We can lower the top index as follows:

$$R_{simj} = g_{sk} R^k{}_{imj}$$

It can be shown that

$$R_{simj} = \Gamma^k{}_{im}\Gamma_{ksj} - \Gamma^k{}_{ij}\Gamma_{ksm} + \frac{\partial \Gamma_{sij}}{\partial u^m} - \frac{\partial \Gamma_{sim}}{\partial u^j} \tag{8.11}$$

By examining Eqs. (8.6) and (8.11), we state some properties of $R^k{}_{imj}$ as follows.

Lemma 8.3.1 *Properties of* ${R^k}_{imj}$:

1. *Antisymmetric property:*

$$\begin{aligned} {R^k}_{imj} &= -{R^k}_{ijm} \\ R_{kijm} &= -R_{kimj} \\ R_{kijm} &= -R_{ikjm} \end{aligned} \tag{8.12}$$

2. *Antisymmetric in the final three indices:*

$${R^k}_{imj} + {R^k}_{jim} + {R^k}_{mji} = 0 \tag{8.13}$$

3. *Block symmetry:*

$$R_{kijm} = R_{jmki} \tag{8.14}$$

4. *From the antisymmetric property:*

$$\begin{aligned} {R^k}_{ijj} &= 0 \\ R_{kijj} &= 0 \end{aligned} \tag{8.15}$$

Exercise 8.3.1

1. Derive the expression of R_{simj} as given in Eq. (8.11).
Solution: We can lower the top index as follows:

$$R_{simj} = g_{sk}{R^k}_{imj}$$

To evaluate the explicit form of R_{limj}, define

$$\Gamma_{sij} = g_{sk}{\Gamma^k}_{ij}$$

and use Eq. (6.14):

$$\frac{\partial g_{sk}}{\partial u^m} = {\Gamma^t}_{sm}g_{tk} + {\Gamma^t}_{km}g_{st} = \Gamma_{ksm} + \Gamma_{skm}$$

$$\begin{aligned} R_{simj} &= g_{sk}{R^k}_{imj} \\ &= g_{sk}{\Gamma^k}_{lm}{\Gamma^l}_{ij} - g_{sk}{\Gamma^k}_{lj}{\Gamma^l}_{im} + g_{sk}\frac{\partial {\Gamma^k}_{ij}}{\partial u^m} - g_{sk}\frac{\partial {\Gamma^k}_{im}}{\partial u^j} \\ &= \Gamma_{slm}{\Gamma^l}_{ij} - \Gamma_{slj}{\Gamma^l}_{im} + g_{sk}\frac{\partial {\Gamma^k}_{ij}}{\partial u^m} - g_{sk}\frac{\partial {\Gamma^k}_{im}}{\partial u^j} \end{aligned}$$

Recall this identity:

$$
\begin{aligned}
\frac{\partial g_{sk}}{\partial u^m} &= \frac{\partial \hat{e}_s \cdot \hat{e}_k}{\partial u^m} \\
&= \frac{\partial \hat{e}_s}{\partial u^m} \cdot \hat{e}_k + \hat{e}_s \cdot \frac{\partial \hat{e}_k}{\partial u^m} \\
&= \Gamma^l{}_{sm} \hat{e}_l \cdot \hat{e}_k + \Gamma^l{}_{km} \hat{e}_l \cdot \hat{e}_s \\
&= \Gamma^l{}_{sm} g_{lk} + \Gamma^l{}_{km} g_{ls} \\
&= \Gamma_{ksm} + \Gamma_{skm}
\end{aligned}
$$

Now, we address the third and fourth terms:

$$
\begin{aligned}
g_{sk} \frac{\partial \Gamma^k{}_{ij}}{\partial u^m} &= \frac{\partial \Gamma_{sij}}{\partial u^m} - \Gamma^k{}_{ij} \frac{\partial g_{sk}}{\partial u^m} \\
&= \frac{\partial \Gamma_{sij}}{\partial u^m} - \Gamma^k{}_{ij} (\Gamma_{ksm} + \Gamma_{skm}) \\
g_{sk} \frac{\partial \Gamma^k{}_{im}}{\partial u^j} &= \frac{\partial \Gamma_{sim}}{\partial u^j} - \Gamma^k{}_{im} (\Gamma_{ksj} + \Gamma_{skj})
\end{aligned}
$$

Putting all the terms together,

$$
\begin{aligned}
R_{simj} &= \Gamma_{slm} \Gamma^l{}_{ij} - \Gamma_{slj} \Gamma^l{}_{im} + \frac{\partial \Gamma_{sij}}{\partial u^m} - \Gamma^k{}_{ij} (\Gamma_{ksm} + \Gamma_{skm}) \\
&\quad - \frac{\partial \Gamma_{sim}}{\partial u^j} + \Gamma^k{}_{im} (\Gamma_{ksj} + \Gamma_{skj})
\end{aligned}
$$

By reindexing all the sums over l by sums over k, we see that some terms cancel:

$$
\begin{aligned}
R_{simj} &= \Gamma_{skm} \Gamma^k{}_{ij} - \Gamma_{skj} \Gamma^k{}_{im} - \Gamma^k{}_{ij} (\Gamma_{ksm} + \Gamma_{skm}) \\
&\quad + \Gamma^k{}_{im} (\Gamma_{ksj} + \Gamma_{skj}) + \frac{\partial \Gamma_{sij}}{\partial u^m} - \frac{\partial \Gamma_{sim}}{\partial u^j} \\
&= \Gamma^k{}_{im} \Gamma_{ksj} - \Gamma^k{}_{ij} \Gamma_{ksm} + \frac{\partial \Gamma_{sij}}{\partial u^m} - \frac{\partial \Gamma_{sim}}{\partial u^j}
\end{aligned}
$$

---

2. Prove the symmetry properties of the Riemann curvature tensor, Eqs. (8.12)–(8.15).

   Solution: First, we prove the antisymmetric property

$$
R^k{}_{imj} = -R^k{}_{ijm}
$$

   Use the Lie bracket representation of the Riemann curvature tensor:

$$
R(X, Y) = [D_X, D_Y] - D_{[X,Y]}
$$

Flipping the two lower indices is equivalent to flipping X, Y in the above equation:

$$\begin{aligned} R(Y,X) &= [D_Y, D_X] - D_{[Y,X]} \\ &= -[D_X, D_Y] - D_{-[X,Y]} \\ &= -([D_X, D_Y] - D_{[X,Y]}) = -R(X,Y) \end{aligned}$$

To prove the antisymmetric property with respect to the lower indices of the Riemann curvature tensor, write in terms of the contraction of the metric tensor:

$$\begin{aligned} R_{kimj} &= g_{tk} R^t{}_{imj} \\ &= -g_{tk} R^t{}_{ijm} \\ &= -R_{kijm} \end{aligned} \tag{8.16}$$

Proving the antisymmetric property by swapping the first two indices is more involved. Use the identity in Eq. (8.16):

$$\frac{\partial g_{si}}{\partial u^m} = \Gamma_{ism} + \Gamma_{sim}$$

Organize the last two terms of the Riemann curvature tensor:

$$R_{simj} = \Gamma^k{}_{im}\Gamma_{ksj} - \Gamma^k{}_{ij}\Gamma_{ksm} + \frac{\partial \Gamma_{sij}}{\partial u^m} - \frac{\partial \Gamma_{sim}}{\partial u^j}$$

We get the intermediate result as

$$\begin{aligned} \frac{\partial \Gamma_{sij}}{\partial u^m} - \frac{\partial \Gamma_{sim}}{\partial u^j} &= \frac{\partial}{\partial u^m}\left(\frac{\partial g_{si}}{\partial u^j} - \Gamma_{isj}\right) - \frac{\partial}{\partial u^j}\left(\frac{\partial g_{si}}{\partial u^m} - \Gamma_{ism}\right) \\ &= -\frac{\partial \Gamma_{isj}}{\partial u^m} + \frac{\partial \Gamma_{ism}}{\partial u^j} \end{aligned}$$

Now, inserting this identity into R_{simj},

$$R_{simj} = \Gamma^k{}_{im}\Gamma_{ksj} - \Gamma^k{}_{ij}\Gamma_{ksm} - \frac{\partial \Gamma_{isj}}{\partial u^m} + \frac{\partial \Gamma_{ism}}{\partial u^j}$$

Now, we are ready to flip the first two indices. We use the identity for raising and lowering indices, $\Gamma^k{}_{sm}\Gamma_{kij} = \Gamma_{ksm}\Gamma^k{}_{ij}$:

$$\begin{aligned} R_{ismj} &= \Gamma_{ksm}\Gamma^k{}_{ij} - \Gamma_{ksj}\Gamma^k{}_{im} - \frac{\partial \Gamma_{sij}}{\partial u^m} + \frac{\partial \Gamma_{sim}}{\partial u^j} \\ &= -R_{simj} \end{aligned}$$

To prove that the sum of cyclic permutations of the last three indices equals zero, simply write out all the terms and note that they cancel each other due to the symmetry properties of the Christoffel symbols. From (8.11),

$$R_{simj} = \Gamma^k{}_{im}\Gamma_{ksj} - \Gamma^k{}_{ij}\Gamma_{ksm} + \frac{\partial \Gamma_{sij}}{\partial u^m} - \frac{\partial \Gamma_{sim}}{\partial u^j}$$

Permute the indices $i \to j$, $m \to i$, $j \to m$:

$$R_{sjim} = \Gamma^k{}_{ji}\Gamma_{ksjm} - \Gamma^k{}_{jm}\Gamma_{ksi} + \frac{\partial \Gamma_{sjm}}{\partial u^i} - \frac{\partial \Gamma_{sji}}{\partial u^m}$$

Permute the indices $i \to m$, $m \to j$, $j \to i$:

$$R_{smji} = \Gamma^k{}_{mj}\Gamma_{ksi} - \Gamma^k{}_{mi}\Gamma_{ksj} + \frac{\partial \Gamma_{smi}}{\partial u^j} - \frac{\partial \Gamma_{smj}}{\partial u^i}$$

Using the symmetric properties of the Christoffel symbols, $\Gamma^k{}_{ij} = \Gamma^k{}_{ji}$, we can see that when we sum the above three equations, all terms cancel one another. Therefore,

$$R_{simj} + R_{sjim} + R_{smji} = 0$$

To prove the block symmetry property,

$$R_{kimj} = R_{mjki}$$

we use the cyclic permutation identity and the antisymmetric properties, $R_{mjki} + R_{mijk} + R_{mkij} = 0$, $R_{kimj} = -R_{kijm}$, $R_{kimj} = R_{ikmj}$:

$$\begin{aligned}
R_{kimj} &= -R_{kjim} - R_{kmji} && \text{(cyclic)} \\
&= R_{jkim} - R_{kmji} && \text{(antisymmetric)} \\
&= -R_{jmki} - R_{jimk} - R_{kmji} && \text{(cyclic)} \\
&= R_{mjki} + R_{jikm} - R_{kmji} && \text{(antisymmetric)}
\end{aligned}$$

We arrive at

$$R_{kimj} - R_{mjki} = R_{jikm} - R_{kmji} \tag{8.17}$$

Reindexing $k \leftrightarrow i$ and $m \leftrightarrow j$,

$$R_{ikjm} - R_{jmik} = R_{mkij} - R_{ijmk}$$

Use antisymmetric properties to swap the indices of the above equation:

$$R_{kimj} - R_{mjki} = R_{kmji} - R_{jikm} \tag{8.18}$$

Add Eqs. (8.17) and (8.18):

$$2(R_{kimj} - R_{mjki}) = 0 \tag{8.19}$$

Therefore, $R_{kimj} = R_{mjki}$.

---

3. Compute the Riemann curvature tensor for:
   a. a 2D sphere
   b. a cylinder

   Solution: To eliminate many terms of ${R^k}_{imj}$, use the convention $u^1 = \theta, u^2 = \phi$ and the following:

   i. ${R^k}_{imm} = 0$, and the first and last columns of the following table contain zeros;
   ii. symmetry identities;
   iii. any partial derivative with respect to u^2 is zero since the Christoffel symbols do not have u^2 terms.

   Now, we can solve the problem quite easily:

   a. Make a table:

$$
\begin{array}{llll}
{R^1}_{111} = 0 & {R^1}_{112} = 0 & {R^1}_{121} = 0 & {R^1}_{122} = 0 \\
{R^1}_{211} = 0 & {R^1}_{212} = -\sin^2\theta & {R^1}_{221} = \sin^2\theta & {R^1}_{222} = 0 \\
{R^2}_{111} = 0 & {R^2}_{112} = 1 & {R^2}_{121} = -1 & {R^2}_{122} = 0 \\
{R^2}_{211} = 0 & {R^2}_{212} = 0 & {R^2}_{221} = 0 & {R^2}_{222} = 0
\end{array}
$$

   b. Then, we need to compute the Christoffel symbols. The parametrization is as follows:

$$
\begin{aligned}
x^1 &= \cos\phi = \cos u^1 \\
x^2 &= \sin\phi = \sin u^1 \\
x^3 &= z = u^2
\end{aligned}
$$

   The tangent vectors are

$$
\begin{aligned}
\hat{e}_1 &= (-\sin u^1, \cos u^1, 0) \\
\hat{e}_2 &= (0, 0, 1) \\
\hat{n} &= (\cos u^1, \sin u^1, 0) \qquad \text{(normal vector)}
\end{aligned}
$$

   The metric tensor:

$$
g_{11} = g_{22} = 1, g_{12} = g_{21} = 0
$$

   Indeed, we can stop here. We see that the metric tensor is the same as that of $\mathbb{R}^2$, whose space is flat. Therefore, all the Christoffel symbols are zero. The Riemann curvature tensor is also zero.

8.4 Jacobi fields

A geodesic is generated by solving the geodesic equation with initial position and tangent vector:

$$\begin{aligned} \frac{d^2\gamma^k}{dt^2} + \Gamma^k{}_{ij}\frac{d\gamma^i}{dt}\frac{d\gamma^j}{dt} &= 0 \\ \gamma(0) &= u_0 \\ \left.\frac{d\gamma}{dt}\right|_{t=0} &= X(0) \in T_uU \end{aligned}$$

Denote the solution of the above geodesic equation by $\gamma(t,0)$. Define a path in the tangent space parameterized by s, $X(s)$. We can solve the geodesic equation with each $X(s)$ to generate a set of geodesics $\gamma(t,s)$. The left panel of Figure 8.3 shows a geodesic generated using $X(0)$, the middle panel shows the uncountable set of geodesics generated by $X(s)$, and the right panel shows the path in tangent space $X(s)$. If we integrate the geodesic equation for each $X(s)$ and connect all geodesics at a synchronized time, say t_1, we can trace a path as shown by the dashed lines in Figure 8.3. This line is specified by $\gamma(t_1,s)$, where t_1 is fixed and s is variable. We can then differentiate with respect to s, yielding the Jacobi field: $\partial\gamma(t_1,s)/\partial s = J(t_1,s)$. Note that the counterpart of $J(t,s)$ is the tangent vectors of the geodesic paths $\dot{\gamma}(t,s)$. Indeed, by this construction, $J(t,s)$ obeys the Jacobi equation

$$D_{\dot{\gamma}}D_{\dot{\gamma}}J - R(\dot{\gamma},J)\dot{\gamma} = 0 \tag{8.20}$$

where $R(\dot{\gamma},J)\dot{\gamma}$ is the Riemann curvature tensor operated on the tangent vectors of the geodesic and the Jacobi vector.

To derive the Jacobi equation, we first state some properties of $J(t,s)$ and $\dot{\gamma}(t,s)$. First, by construction, it is obvious that the flow of $J(t,s)$ and the flow of $\dot{\gamma}(t,s)$ commute. That is,

$$\sigma_{J,s} \circ \sigma_{\dot{\gamma},t} = \sigma_{\dot{\gamma},t} \circ \sigma_{J,s}$$

Another way to represent this is

$$[\dot{\gamma}, J] = 0$$

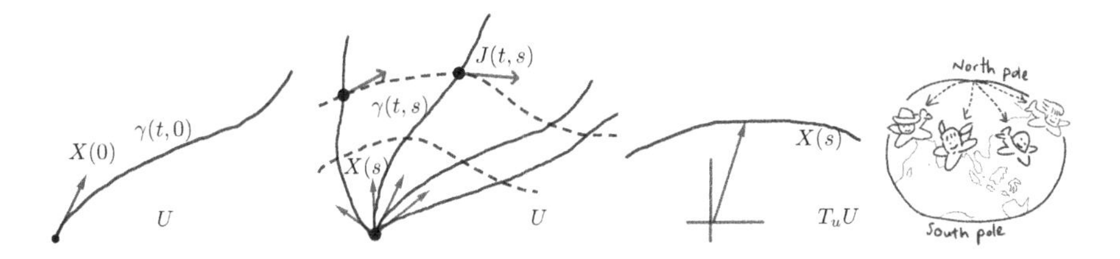

Figure 8.3: The left panel shows a geodesic $\gamma(t, 0)$ generated by a tangent vector $X(0)$. The middle panel shows a set of geodesics $\gamma(t, s)$ generated by a set of initial tangent vectors $X(s)$. Tracing along the time component t and connecting all the geodesic points at equivalent times will create a level-set curve indicated by dashed lines. The tangent vector fields along the dashed lines are called the Jacobi fields $J(t, s)$. The right panel shows a path on the tangent space $T_u U$ with a path specifying $X(s)$. On the far right is a cartoon illustration of the Jacobi fields of Earth.

Using the torsion-free property of the covariant derivative,

$$D_{\dot{\gamma}} J - D_J \dot{\gamma} = [\dot{\gamma}, J] = 0$$

Also, note that

$$[D_{\dot{\gamma}}, D_J]\dot{\gamma} = R(\dot{\gamma}, J)\dot{\gamma} = D_{\dot{\gamma}} D_J \dot{\gamma} - D_J D_{\dot{\gamma}} \dot{\gamma}$$

Since γ is a geodesic, $D_{\dot{\gamma}}\dot{\gamma} = 0$. Then, $D_J D_{\dot{\gamma}} \dot{\gamma} = 0$, so

$$D_{\dot{\gamma}} D_J \dot{\gamma} - R(\dot{\gamma}, J)\dot{\gamma} = 0$$

Lastly, swapping J and $\dot{\gamma}$, $D_J \dot{\gamma} = D_{\dot{\gamma}} J$, which leads to the Jacobi equation:

$$D_{\dot{\gamma}} D_{\dot{\gamma}} J - R(\dot{\gamma}, J)\dot{\gamma} = 0$$

We are now in a position to ask, "What does the Jacobi equation tell us?" The first term is a second derivative of J, which is the acceleration of the change in J (both in magnitude and direction). Recall that J represents the density of separation of geodesic paths as we increase the integration time t. Hence, the first term is the rate of change of separation of geodesics. Many authors call this the geodesic deviation. The Jacobi equation tells us that the rate of change of geodesic deviation depends on the Riemann curvature. Hence, for a flat space, where $R = 0$, we have

$$D_{\dot{\gamma}} J = \text{constant vector}$$

Figure 8.4 illustrates that the distance of two simultaneous points tracing out geodesics in $\mathbb{R}^2$ changes at a constant rate.

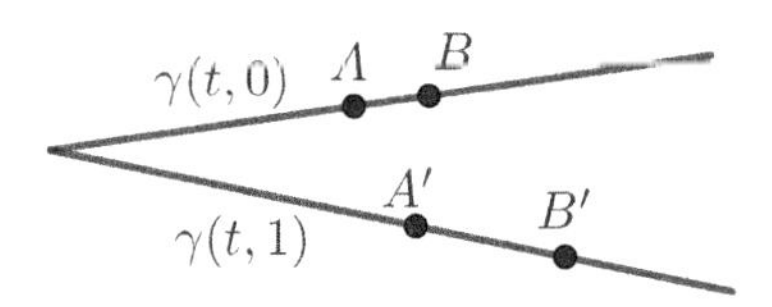

Figure 8.4: An illustration of the idea that the distance between two simultaneous points (A and A', B and B') tracing out straight lines ($\gamma(t,0)$ and $\gamma(t,1)$) in $\mathbb{R}^2$ increases at a constant rate.

Finally, the expression of the Riemann curvature tensor $R(\dot{\gamma}, J)\dot{\gamma}$ is interesting. It measures the parallel transport of $\dot{\gamma}$ along two paths: the first path along $\dot{\gamma}$ and then along J; the second path along J and then along $\dot{\gamma}$. This term does not parallel transport any arbitrary vectors, but it transports $\dot{\gamma}$ itself. Indeed, this expression is important for deriving the Ricci tensor through the use of the sectional curvature. We elaborate upon the sectional curvature and Ricci tensor in the following section.

8.5 Sectional Curvature, the Ricci Tensor, and Scalar Curvature

8.5.1 Sectional curvature

Recall that in the previous section, we obtained the Jacobi equation, which is given by,

$$D_{\dot{\gamma}} D_{\dot{\gamma}} J - R(\dot{\gamma}, J)\dot{\gamma} = 0$$

where we have an expression of the form $R(X,Y)X$. In this section, we explore the expression $R(X,Y)X \cdot Y$, which is the change of the vector X under parallel transport along a parallelogram X, Y and projected to Y. This quantity should have something to do with the geodesic deviation. Next, we define the sectional curvature as

$$K(X,Y) = \frac{R(X,Y)X \cdot Y}{(X \cdot X)(Y \cdot Y) - (X \cdot Y)^2} \tag{8.21}$$

where $(X \cdot X)(Y \cdot Y) - (X \cdot Y)^2$ is the square of the area of the parallelogram formed by X and Y. We leave it as an exercise for the reader to show this. $K(X,Y)$ is coordinate independent, and we leave it as an exercise for the reader to proof this fact. Since the sectional curvature is coordinate independent, one can always find a orthonormal coordinate ∂_{u^i} such that

$$K(X,Y) = K(\partial_{u^i}, \partial_{u^j}) = R(\partial_{u^i}, \partial_{u^j})\partial_{u^i} \cdot \partial_{u^j}$$

8.5.2 Ricci tensor

Given a point on the manifold $u_0 \in U$, construct an orthogonal basis on $T_u U$, $\partial_{u^1}, \dots, \partial_{u^d}$. We define the Ricci curvature as

$$Ric(\partial_{u^d}, \partial_{u^d}) = \frac{1}{d-1} \sum_{i}^{d-1} R(\partial_{u^d}, \partial_{u^i}) \partial_{u^d} \cdot \partial_{u^i} \in \mathbb{R} \tag{8.22}$$

We can see that we can construct geodesics along ∂_{u^d}. We find the geodesic deviation along the ∂_{u^i} direction using $R(\partial_{u^d}, \partial_{u^i})\partial_{u^d}$ and then project onto ∂_{u^i}. Perform this operation in all directions and then take the average. Note that the Ricci tensor is a scalar, is evaluated at a point, and takes in one tangent vector. It is often interpreted as the rate of change of a volume element along a direction. Next, we write out the components of the Ricci tensor:

$$Ric_{dd} = \frac{1}{d-1} \sum_{i} R^k{}_{ddi} \delta_{ki} = \frac{1}{d-1} R^i{}_{ddi}$$

In some literature, the components of the Ricci tensor are defined as:

$$Ric_{lm} = R^i{}_{lim} \tag{8.23}$$

which we observe to be different from our equation by the factor of $1/(d-1)$ and a minus sign. The minus sign comes from the sign convention of the Riemann curvature tensor. The Ricci tensor can also be defined to take in two different tangent vectors:

$$Ric(X, Y) = \frac{1}{d} \sum_{i}^{d} R(X, \partial_{u^i}) Y \cdot \partial_{u^i} \in \mathbb{R} \tag{8.24}$$

8.5.3 Scalar curvature

The Ricci curvature computes the average of geodesic deviations in a certain direction. The scalar curvature takes the average of the Ricci tensor in all directions:

$$S = \frac{1}{d} \sum_{i}^{d} Ric(\partial_{u^i}, \partial_{u^i}) \tag{8.25}$$

In terms of components, the scalar curvature is given by

$$S = g^{ij} R_{ij} \tag{8.26}$$

8.5.4 Schwarzschild solution for non-rotating black holes

We end this chapter with an exercise to compute the event horizon (the Schwarzschild radius) of a non-rotating black hole. The objective is to compute the metric tensor and look for singularities in it. Since the metric tensor is unknown, we write it in the form of unknown functions, $U(r)$ and $V(r)$, where r is the radial distance from the black hole in space-time spherical coordinates. From the metric tensor (with unknown functions), we derive the Christoffel symbols and hence the Riemann curvature tensor. From the Riemann curvature tensor, we compute the Ricci tensor and the Ricci scalar all in terms of the unknown functions $U(r)$ and $V(r)$. Finally, we use the Einstein field equations to find the solutions of $U(r)$ and $V(r)$.

Exercise 8.5.1

Schwarzschild solution for a non-rotating black hole

In this exercise, we derive the Schwarzschild solution for the event horizon of a non-rotating black hole in vacuum. We break down this derivation into

Table 8.1: Non-zero Christoffel symbols.

Upper index	Lower indices	value
0	01,10	${\Gamma^0}_{10} = {\Gamma^0}_{01} = \frac{U'}{2U}$
1	00	${\Gamma^1}_{00} = \frac{U'}{2V}$
	11	${\Gamma^1}_{11} = \frac{V'}{2V}$
	22	${\Gamma^1}_{22} = -\frac{r}{V}$
	33	${\Gamma^1}_{33} = -\frac{r\sin^2\theta}{V}$
2	12,21	${\Gamma^2}_{21} = {\Gamma^2}_{12} = \frac{1}{r}$
	33	${\Gamma^2}_{33} = -\sin\theta\cos\theta$
3	13,31	${\Gamma^3}_{13} = {\Gamma^3}_{31} = \frac{1}{r}$
	23,32	${\Gamma^3}_{23} = {\Gamma^3}_{32} = \cot\theta$

smaller parts to help the reader follow and understand the process. First, we list some claims and assumptions:

> [Assumption] spherical symmetry;
>
> [Assumption] there is no dynamics in the geometry, i.e. the metric tensor does not change with time;
>
> [Assumption] since the black hole is in vacuum, $T_{ij} = 0$;
>
> [Assumption] the cosmological constant is small compared to other terms since we are not working at cosmological scales, and we neglect it in our solution, i.e. $\Lambda = 0$;
>
> [Claim] the metric tensor has zeroes as its off-diagonal elements, which we claim without a proof.

1. The Minkowski metric in the Cartesian coordinate representation $(\hat{u}^0, \hat{u}^1, \hat{u}^2, \hat{u}^3) = (t, x, y, z)$ in vacuum far away from any mass is given by
$$\hat{g}_{ij} = \delta_{ij}(\delta_{0i} - \delta_{1i} - \delta_{2i} - \delta_{3i}) \tag{8.27}$$
Using the pullback equation
$$g_u(\partial_{u^i}, \partial_{u^j}) = \hat{g}_{\varphi(u)}(\varphi_* \partial_{u^i}, \varphi_* \partial_{u^j})$$
show that the Minkowski metric in polar coordinate representation $(u^0, u^1, u^2, u^3) = (t, r, \theta, \phi)$ far away from any mass is given by
$$g = \begin{pmatrix} 1 & 0 & 0 & 0 \\ 0 & -1 & 0 & 0 \\ 0 & 0 & -r^2 & 0 \\ 0 & 0 & 0 & -r^2 \sin^2 \theta \end{pmatrix}$$
The mapping φ is given by spherical coordinates:
$$\begin{aligned} t = \hat{u}^0 &= u^0 = t \\ x = \hat{u}^1 &= u^1 \sin u^2 \cos u^3 = r \sin\theta \cos\phi \\ y = \hat{u}^2 &= u^1 \sin u^2 \sin u^3 = r \sin\theta \sin\phi \\ z = \hat{u}^3 &= u^1 \cos u^2 = r\cos\theta \end{aligned}$$
2. Write down the functional form of the metric tensor close to the black hole.
3. Show that the Christoffel symbols are zero except for the components listed in Table 8.1.

4. The Einstein field equation is given by

$$Ric_{ij} - \frac{1}{2} S g_{ij} + \Lambda g_{ij} = \frac{8\pi G}{c^4} T_{ij} \tag{8.28}$$

The scalar curvature is given by $S = g^{ij} R_{ij}$. Using $\Lambda = 0$ and $T_{ij} = 0$, show that the scalar curvature $S = 0$ and hence the Ricci tensor $Ric_{ij} = 0$.

5. If we let $g_{00} = U$ and $g_{11} = V$, show that the Ricci tensor in terms of the metric tensor is given by

$$Ric = \begin{pmatrix} \begin{matrix} \frac{U''}{2V} + \frac{U'}{rV} \\ -\frac{U'V'}{4V^2} - \frac{U'^2}{4UV} \end{matrix} & 0 & 0 & 0 \\ 0 & \begin{matrix} -\frac{U''}{2U} + \frac{U'V'}{4UV} \\ +\frac{U'^2}{4U^2} + \frac{V'}{rV} \end{matrix} & 0 & 0 \\ 0 & 0 & \begin{matrix} -\frac{rU'}{2UV} - \frac{1}{V} \\ +\frac{rV'}{2V^2} + 1 \end{matrix} & 0 \\ 0 & 0 & 0 & \left(-\frac{rU'}{2UV} + \frac{rV'}{2V^2} - \frac{1}{V} + 1\right) \sin^2\theta \end{pmatrix} \tag{8.29}$$

6. The metric tensor has a singularity at the event horizon (or the Schwarzschild radius). Show that it is given by

$$r_s \approx \frac{2GM}{c^2}$$

where G is the gravitational constant, M is the mass of the black hole, and c is the speed of light in vacuum.

Solution:

Pullback of metric tensor

To obtain the metric tensor far away from any mass in spherical coordinates, we need to perform pushforward. Write the tangent vector as

$$\partial_{u^i} = \delta^k{}_i \partial_{u^k} = X^k \partial_{u^k}$$

So, $X^k = \delta^k{}_i$. The pushforward equation for the components is

$$\begin{aligned} \hat{X}^j &= \frac{\partial \varphi^j}{\partial u^k} X^k \\ &= \frac{\partial \varphi^j}{\partial u^k} \delta^k{}_i \\ &= \frac{\partial \varphi^j}{\partial u^i} \end{aligned}$$

The pushforward vector is then

$$\varphi_* \partial_{u^i} = \hat{X}^j \partial_{\hat{u}^j} = \frac{\partial \varphi^j}{\partial u^i} \partial_{\hat{u}^j}$$

For $i = 0$,

$$\varphi_* \partial_{u^0} = \partial_{\hat{u}^0}$$

For $i = 1$,

$$\varphi_* \partial_{u^1} = \sin\theta \cos\phi \partial_{\hat{u}^1} + \sin\theta \sin\phi \partial_{\hat{u}^2} + \cos\theta \partial_{\hat{u}^3}$$

For $i = 2$,

$$\varphi_* \partial_{u^2} = r \cos\theta \cos\phi \partial_{\hat{u}^1} + r \cos\theta \sin\phi \partial_{\hat{u}^2} - r \sin\theta \partial_{\hat{u}^3}$$

For $i = 3$,

$$\varphi_* \partial_{u^3} = -r \sin\theta \sin\phi \partial_{\hat{u}^1} + r \sin\theta \cos\phi \partial_{\hat{u}^2}$$

Now, we are ready to compute the pullback of the metric tensor using the Minkowski metric given in Eq. (8.27). For $i = j = 0$,

$$g(\partial_{u^0}, \partial_{u^0}) = \hat{g}(\varphi_* \partial_{u^0}, \varphi_* \partial_{u^0}) = \hat{g}(\partial_{\hat{u}^0}, \partial_{\hat{u}^0}) = 1$$

For $i = 0, j \neq 0$,

$$g(\partial_{u^0}, \partial_{u^j}) = \hat{g}(\varphi_* \partial_{u^0}, \varphi_* \partial_{u^j}) = \hat{g}(\partial_{\hat{u}^0}, \varphi_* \partial_{u^j}) = 0$$

Because $\hat{g}(\partial_{\hat{u}^0}, \partial_{\hat{u}^j}) = 0$ for $j \neq 0$.
For $i = 1, j = 1$,

$$\begin{aligned} g(\partial_{u^1}, \partial_{u^1}) &= \hat{g}(\varphi_* \partial_{u^1}, \varphi_* \partial_{u^1}) \\ &= -(\sin\theta \cos\phi)^2 - (\sin\theta \sin\phi)^2 - (\cos\theta)^2 = -1 \end{aligned}$$

For $i = 1, j = 2$,

$$\begin{aligned} g(\partial_{u^1}, \partial_{u^2}) = & -r(\cos\theta \cos\phi)(\sin\theta \cos\phi) - r(\cos\theta \sin\phi)(\sin\theta \sin\phi) \\ & + r(\sin\theta \cos\theta) = 0 \end{aligned}$$

For $i = 1, j = 3$,

$$g(\partial_{u^1}, \partial_{u^3}) = r(\sin\theta \sin\phi)(\sin\theta \cos\phi) - r(\sin\theta \cos\phi)(\sin\theta \sin\phi) = 0$$

For $i = 2, j = 3$,

$$g(\partial_{u^2}, \partial_{u^3}) = r^2(\sin\theta \sin\phi)(\cos\theta \cos\phi) - r^2(\sin\theta \cos\phi)(\cos\theta \sin\phi) = 0$$

By symmetry,

$$\begin{aligned} g(\partial_{u^1}, \partial_{u^2}) &= g(\partial_{u^2}, \partial_{u^1}) = 0 \\ g(\partial_{u^2}, \partial_{u^3}) &= g(\partial_{u^3}, \partial_{u^2}) = 0 \\ g(\partial_{u^1}, \partial_{u^3}) &= g(\partial_{u^3}, \partial_{u^1}) = 0 \end{aligned}$$

For $i = 2, j = 2$,

$$g(\partial_{u^2}, \partial_{u^2}) = -r^2(\cos\theta\cos\phi)^2 - r^2(\cos\theta\sin\phi)^2 - r^2(\sin\theta)^2 = -r^2$$

For $i = 3, j = 3$,

$$g(\partial_{u^3}, \partial_{u^3}) = -r^2(\sin\theta\sin\phi)^2 - r^2(\sin\theta\cos\phi)^2 = -r^2\sin^2\theta$$

In matrix form,

$$g = \begin{pmatrix} 1 & 0 & 0 & 0 \\ 0 & -1 & 0 & 0 \\ 0 & 0 & -r^2 & 0 \\ 0 & 0 & 0 & -r^2\sin^2\theta \end{pmatrix}$$

Functional form of metric tensor nearer black hole

The above metric tensor is for points far away from the black hole. For points nearer the black hole, the metric tensor must depend on their distances from the black hole. By spherical symmetry, there should not be any angle dependence. Also, using our claim (without proof) that off-diagonal elements are zeroes, we write

$$g = \begin{pmatrix} U(r) & 0 & 0 & 0 \\ 0 & -V(r) & 0 & 0 \\ 0 & 0 & -r^2 & 0 \\ 0 & 0 & 0 & -r^2\sin^2\theta \end{pmatrix}$$

Christoffel symbols

From the metric tensor, we can derive the Christoffel symbols. From the Christoffel symbols, we can compute the Riemann curvature and then the Ricci tensor:

$$\Gamma^k{}_{ij} = \frac{1}{2}g^{ks}\left(\frac{\partial g_{si}}{\partial u^j} + \frac{\partial g_{sj}}{\partial u^i} - \frac{\partial g_{ij}}{\partial u^s}\right)$$

Express the derivatives of the metric tensor as

$$\frac{\partial g}{\partial t} = \frac{\partial g}{\partial u^0} = \begin{pmatrix} 0 & 0 & 0 & 0 \\ 0 & 0 & 0 & 0 \\ 0 & 0 & 0 & 0 \\ 0 & 0 & 0 & 0 \end{pmatrix}, \qquad \partial_t g_{ij} = 0,$$

$$\frac{\partial g}{\partial r} = \frac{\partial g}{\partial u^1} = \begin{pmatrix} \frac{\partial U}{\partial r} & 0 & 0 & 0 \\ 0 & -\frac{\partial V}{\partial r} & 0 & 0 \\ 0 & 0 & -2r & 0 \\ 0 & 0 & 0 & -2r\sin^2\theta \end{pmatrix},$$

$$\frac{\partial g}{\partial \theta} = \frac{\partial g}{\partial u^2} = \begin{pmatrix} 0 & 0 & 0 & 0 \\ 0 & 0 & 0 & 0 \\ 0 & 0 & 0 & 0 \\ 0 & 0 & 0 & -2r^2\sin\theta\cos\theta \end{pmatrix},$$

$$\frac{\partial g}{\partial \phi} = \frac{\partial g}{\partial u^3} = \begin{pmatrix} 0 & 0 & 0 & 0 \\ 0 & 0 & 0 & 0 \\ 0 & 0 & 0 & 0 \\ 0 & 0 & 0 & 0 \end{pmatrix}, \qquad \partial_\phi g_{ij} = 0$$

For ${\Gamma^0}_{ij}$,

$${\Gamma^0}_{ij} = \frac{1}{2} g^{0s} \left(\frac{\partial g_{is}}{\partial u^j} + \frac{\partial g_{js}}{\partial u^i} - \frac{\partial g_{ij}}{\partial u^s} \right)$$

Using $g^{0s} = \delta^{0s}/U$, $\partial g_{ij}/\partial u^0 = 0$.

$$\begin{aligned} {\Gamma^0}_{ij} &= \frac{1}{2U} \left(\frac{\partial g_{i0}}{\partial u^j} + \frac{\partial g_{j0}}{\partial u^i} - \frac{\partial g_{ij}}{\partial u^0} \right) \qquad \text{(using } \partial_{u^0} g_{ij} = 0\text{)} \\ &= \frac{1}{2U} \left(\frac{\partial g_{i0}}{\partial u^j} + \frac{\partial g_{j0}}{\partial u^i} \right) \end{aligned}$$

For ${\Gamma^0}_{00}$,

$${\Gamma^0}_{00} = \frac{1}{2U} \left(\frac{\partial g_{00}}{\partial t} + \frac{\partial g_{00}}{\partial t} \right) = 0$$

For ${\Gamma^0}_{01}$,

$${\Gamma^0}_{01} = \frac{1}{2U} \left(\frac{\partial g_{00}}{\partial r} + \frac{\partial g_{10}}{\partial t} \right) = \frac{1}{2U} \frac{\partial U}{\partial r} = \frac{U'}{2U}$$

For ${\Gamma^0}_{02}$,

$${\Gamma^0}_{02} = \frac{1}{2U}\left(\frac{\partial g_{00}}{\partial \theta} + \frac{\partial g_{20}}{\partial t}\right) = 0$$

For ${\Gamma^0}_{03}$,

$${\Gamma^0}_{03} = \frac{1}{2U}\left(\frac{\partial g_{00}}{\partial \phi} + \frac{\partial g_{30}}{\partial t}\right) = 0$$

For ${\Gamma^0}_{ij}$, $i \neq 0$ and $j \neq 0$, using $g_{0i} = 0$ for $i \neq 0$,

$${\Gamma^0}_{ij} = \frac{1}{2U}\left(\frac{\partial g_{0i}}{\partial u^j} + \frac{\partial g_{j0}}{\partial u^i}\right) = 0$$

Using the same trick, $g^{1s} = -\delta^{1s}/V$, we work out the Christoffel symbol for the upper index of 1. For ${\Gamma^1}_{ij}$,

$$\begin{aligned}
{\Gamma^1}_{ij} &= \frac{1}{2}g^{1s}\left(\frac{\partial g_{si}}{\partial u^j} + \frac{\partial g_{sj}}{\partial u^i} - \frac{\partial g_{ij}}{\partial u^s}\right) \\
&= -\frac{1}{2V}\delta^{1s}\left(\frac{\partial g_{si}}{\partial u^j} + \frac{\partial g_{sj}}{\partial u^i} - \frac{\partial g_{ij}}{\partial u^s}\right) \\
&= -\frac{1}{2V}\left(\frac{\partial g_{1i}}{\partial u^j} + \frac{\partial g_{1j}}{\partial u^i} - \frac{\partial g_{ij}}{\partial r}\right)
\end{aligned}$$

For ${\Gamma^1}_{00}$,

$${\Gamma^1}_{00} = \frac{1}{2V}\frac{\partial U}{\partial r} = \frac{U'}{2V}$$

For ${\Gamma^1}_{0j}$, $j \neq 0$,

$${\Gamma^1}_{0j} = -\frac{1}{2V}\left(\frac{\partial g_{10}}{\partial u^j} + \frac{\partial g_{1j}}{\partial t}\right) = 0$$

For ${\Gamma^1}_{11}$,

$${\Gamma^1}_{11} = -\frac{1}{2V}\left(\frac{\partial g_{11}}{\partial r} + \frac{\partial g_{11}}{\partial r} - \frac{\partial g_{11}}{\partial r}\right) = \frac{V'}{2V}$$

For ${\Gamma^1}_{12}$,

$${\Gamma^1}_{12} = -\frac{1}{2V}\left(\frac{\partial g_{11}}{\partial \theta} + \frac{\partial g_{12}}{\partial r} - \frac{\partial g_{12}}{\partial r}\right) = 0$$

Similarly, ${\Gamma^1}_{13} = {\Gamma^1}_{23} = 0$.

For ${\Gamma^1}_{22}$,

$$ {\Gamma^1}_{22} = -\frac{1}{2V}\left(\frac{\partial g_{12}}{\partial \theta} + \frac{\partial g_{12}}{\partial \theta} - \frac{\partial g_{22}}{\partial r}\right) = -\frac{1}{2V}\frac{\partial r^2}{\partial r} = -\frac{r}{V} $$

For ${\Gamma^1}_{33}$,

$$ {\Gamma^1}_{33} = -\frac{1}{2V}\left(\frac{\partial g_{13}}{\partial \phi} + \frac{\partial g_{13}}{\partial \phi} - \frac{\partial g_{33}}{\partial r}\right) = -\frac{1}{2V}\frac{\partial r^2 \sin^2\theta}{\partial r} = -\frac{r\sin^2\theta}{V} $$

Using the same trick, $g^{2s} = -\delta^{2s}/r^2$, we work out the Christoffel symbol for the upper index of 2. For ${\Gamma^2}_{ij}$,

$$ {\Gamma^2}_{ij} = -\frac{1}{2r^2}\left(\frac{\partial g_{2i}}{\partial u^j} + \frac{\partial g_{2j}}{\partial u^i} - \frac{\partial g_{ij}}{\partial \theta}\right) $$

For ${\Gamma^2}_{0j}$,

$$ {\Gamma^2}_{0j} = -\frac{1}{2r^2}\left(\frac{\partial g_{20}}{\partial u^j} + \frac{\partial g_{2j}}{\partial t} - \frac{\partial g_{0j}}{\partial \theta}\right) = 0 $$

For ${\Gamma^2}_{11}$,

$$ {\Gamma^2}_{11} = -\frac{1}{2r^2}\left(\frac{\partial g_{21}}{\partial r} + \frac{\partial g_{21}}{\partial r} - \frac{\partial g_{11}}{\partial \theta}\right) = 0 $$

For ${\Gamma^2}_{12}$,

$$ {\Gamma^2}_{12} = -\frac{1}{2r^2}\left(\frac{\partial g_{21}}{\partial \theta} + \frac{\partial g_{22}}{\partial r} - \frac{\partial g_{12}}{\partial \theta}\right) = -\frac{1}{2r^2}\frac{\partial(-r^2)}{\partial r} = \frac{1}{r} $$

For ${\Gamma^2}_{13} = 0$.
For ${\Gamma^2}_{22}$,

$$ {\Gamma^2}_{22} = -\frac{1}{2r^2}\left(\frac{\partial g_{22}}{\partial \theta} + \frac{\partial g_{22}}{\partial \theta} - \frac{\partial g_{22}}{\partial \theta}\right) = 0 $$

For ${\Gamma^2}_{23} = 0$.
For ${\Gamma^2}_{33}$,

$$ {\Gamma^2}_{33} = -\frac{1}{2r^2}\left(\frac{\partial g_{23}}{\partial \phi} + \frac{\partial g_{23}}{\partial \phi} - \frac{\partial g_{33}}{\partial \theta}\right) = -\sin\theta\cos\theta $$

Next, we compute the Christoffel symbol for the upper index of 3 using $g^{3s} = -\delta^{3s}/r^2 \sin^2\theta$:

$$\begin{aligned} \Gamma^3{}_{ij} &= -\frac{1}{2r^2\sin^2\theta}\left(\frac{\partial g_{3i}}{\partial u^j} + \frac{\partial g_{3j}}{\partial u^i} - \frac{\partial g_{ij}}{\partial \phi}\right) \\ &= -\frac{1}{2r^2\sin^2\theta}\left(\frac{\partial g_{3i}}{\partial u^j} + \frac{\partial g_{3j}}{\partial u^i}\right) \end{aligned}$$

The only non-zero terms are either $i = 3$ or $j = 3$. By the symmetric property of the Christoffel symbols, we consider only $\Gamma^3{}_{30}$, $\Gamma^3{}_{31}$, $\Gamma^3{}_{32}$, and $\Gamma^3{}_{33}$. $\Gamma^3{}_{30} = 0$ and $\Gamma^3{}_{33} = 0$ because differentials with t or ϕ are zero. The only non-zero terms are $\Gamma^3{}_{31}$ and $\Gamma^3{}_{32}$:

$$\Gamma^3{}_{31} = -\frac{1}{2r^2\sin^2\theta}\left(\frac{\partial g_{33}}{\partial r} + \frac{\partial g_{31}}{\partial \phi}\right) = -\frac{1}{2r^2\sin^2\theta}\frac{\partial(-r^2\sin^2\theta)}{\partial r} = \frac{1}{r}$$

$$\Gamma^3{}_{32} = -\frac{1}{2r^2\sin^2\theta}\left(\frac{\partial g_{33}}{\partial \theta} + \frac{\partial g_{32}}{\partial \phi}\right) = -\frac{1}{2r^2\sin^2\theta}\frac{\partial(-r^2\sin^2\theta)}{\partial \theta} = \cot\theta$$

Collect all the non-zero terms, as shown in Table 8.1.

Ricci scalar

The Ricci scalar is a contraction of the Ricci tensor. First, raise one index and then sum over:

$$S = g^{ik} Ric_{ik}$$

From the Einstein field equations, we can perform a contraction:

$$Ric_{ij} - \frac{1}{2} S g_{ij} = 0$$

Contract both sides by g^{ij}:

$$\begin{aligned} Ric_{ij}g^{ij} - \frac{1}{2} S g_{ij} g^{ij} &= 0 \\ S - \frac{1}{2} S \delta^i{}_i &= 0 \\ S - \frac{1}{2} 4S &= 0 \\ -S &= 0 \end{aligned}$$

So, substituting $S = 0$ back into the Einstein field equations,

$$Ric_{ij} = 0$$

Ricci tensor

We stated without proof that the form of the metric tensor is a diagonal matrix, that is, $g_{ij} = 0$, for $i \neq j$. Using the Einstein field equations, $Ric_{ij} = 0$. We evaluate Ric_{00}, Ric_{11}, Ric_{22}, and Ric_{33} to solve for $U(r)$ and $V(r)$.
The Ricci tensor is a contraction of the Riemann curvature tensor:

$$Ric_{ij} = R^k{}_{ikj} = \Gamma^k{}_{lek}\Gamma^l{}_{ij} - \Gamma^k{}_{lj}\Gamma^l{}_{ik} + \frac{\partial \Gamma^k{}_{ij}}{\partial u^k} - \frac{\partial \Gamma^k{}_{ik}}{\partial u^j}$$

The Einstein summation convention is applied to the indices k and l. We first work out some identities which would be useful for later computations. By direction substitutions,

$$\Gamma^k{}_{lek} = \Gamma^0{}_{l0} + \Gamma^1{}_{l1} + \Gamma^2{}_{l2} + \Gamma^3{}_{l3}$$

$$\begin{aligned}
\Gamma^k{}_{0k} &= \Gamma^0{}_{00} + \Gamma^1{}_{01} + \Gamma^2{}_{02} + \Gamma^3{}_{03} \\
&= 0 + 0 + 0 + 0 = 0 \\
\Gamma^k{}_{1k} &= \Gamma^0{}_{10} + \Gamma^1{}_{11} + \Gamma^2{}_{12} + \Gamma^3{}_{13} \\
&= \frac{U'}{2U} + \frac{V'}{2V} + \frac{1}{r} + \frac{1}{r} = W \\
\Gamma^k{}_{2k} &= \Gamma^0{}_{20} + \Gamma^1{}_{21} + \Gamma^2{}_{22} + \Gamma^3{}_{23} \\
&= 0 + 0 + 0 + \cot\theta = \cot\theta \\
\Gamma^k{}_{3k} &= \Gamma^0{}_{30} + \Gamma^1{}_{31} + \Gamma^2{}_{32} + \Gamma^3{}_{33} \\
&= 0 + 0 + 0 + 0 = 0
\end{aligned} \tag{8.30}$$

For Ric_{00}, evaluate term by term,

$$Ric_{00} = A_{00} - B_{00} + C_{00} - D_{00} = \Gamma^k{}_{lek}\Gamma^l{}_{00} - \Gamma^k{}_{l0}\Gamma^l{}_{0k} + \frac{\partial \Gamma^k{}_{00}}{\partial u^k} - \frac{\partial \Gamma^k{}_{0k}}{\partial t}$$

From the table of non-zero Christoffel symbols, the first term $\Gamma^k{}_{lk}\Gamma^l{}_{00} \neq 0$ only when $l = 1$. Also, use $\Gamma^k{}_{1k} = W$. So, the first term is

$$A_{00} = \Gamma^k{}_{1k}\Gamma^1{}_{00} = \frac{WU'}{2V} = \left(\frac{U'}{2U} + \frac{V'}{2V} + \frac{2}{r}\right)\frac{U'}{2V}$$

The second term is non-zero only when $l = 0, 1$:

$$B_{00} = \Gamma^k{}_{l0}\Gamma^l{}_{0k} = \Gamma^k{}_{00}\Gamma^0{}_{0k} + \Gamma^k{}_{10}\Gamma^1{}_{0k}$$

Use the lookup table of the non-zero Christoffel symbols again to evaluate the above equation. We get terms of ${\Gamma^k}_{00} \neq 0$ only when $k = 1$ and ${\Gamma^k}_{10} \neq 0$ only when $k = 0$:

$$B_{00} = {\Gamma^k}_{l0}{\Gamma^l}_{0k} = {\Gamma^1}_{00}{\Gamma^0}_{01} + {\Gamma^0}_{10}{\Gamma^1}_{00} = 2{\Gamma^1}_{00}{\Gamma^0}_{01} = \frac{U'^2}{2UV}$$

For C_{00}, the only non-zero term is when $k = 1$:

$$\frac{\partial {\Gamma^1}_{00}}{\partial r} = \frac{\partial}{\partial r}\left(\frac{U'}{2V}\right) = \frac{U''}{2V} - \frac{U'V'}{2V^2}$$

$D_{00} = 0$ because it is a differential with respect to t, and the Christoffel symbols do not depend on t. Therefore, we have

$$\begin{aligned} Ric_{00} &= \frac{U'^2}{4UV} + \frac{U'V'}{4V^2} + \frac{U'}{rV} - \frac{U'^2}{2UV} + \frac{U''}{2V} - \frac{U'V'}{2V^2} \\ &= \frac{U''}{2V} + \frac{U'}{rV} - \frac{U'^2}{4UV} - \frac{U'V'}{4V^2} \end{aligned} \tag{8.31}$$

The next component is Ric_{11}:

$$Ric_{11} = A_{11} - B_{11} + C_{11} - D_{11} = {\Gamma^k}_{lek}{\Gamma^l}_{11} - {\Gamma^k}_{l1}{\Gamma^l}_{1k} + \frac{\partial {\Gamma^k}_{11}}{\partial u^k} - \frac{\partial {\Gamma^k}_{1k}}{\partial r}$$

The first term is non-zero only when $l = 1$:

$$A_{11} = {\Gamma^k}_{1k}{\Gamma^1}_{11} = \left(\frac{U'}{2U} + \frac{V'}{2V} + \frac{2}{r}\right)\frac{V'}{2V}$$

B_{11} is non-zero when $l = 0, 1, 2, 3$. Expanding the four terms,

$${\Gamma^k}_{l1}{\Gamma^l}_{1k} = {\Gamma^k}_{01}{\Gamma^0}_{1k} + {\Gamma^k}_{11}{\Gamma^1}_{1k} + {\Gamma^k}_{21}{\Gamma^2}_{1k} + {\Gamma^k}_{31}{\Gamma^3}_{1k}$$

Use the table of the non-zero Christoffel symbols to match the upper indices:

$$\begin{aligned} B_{11} = {\Gamma^k}_{l1}{\Gamma^l}_{1k} &= {\Gamma^0}_{01}{\Gamma^0}_{10} + {\Gamma^1}_{11}{\Gamma^1}_{11} + {\Gamma^2}_{21}{\Gamma^2}_{12} + {\Gamma^3}_{31}{\Gamma^3}_{13} \\ &= \frac{U'^2}{4U^2} + \frac{V'^2}{4V^2} + \frac{2}{r^2} \end{aligned}$$

In C_{11}, the only non-zero term is when $l = 1$:

$$C_{11} = \frac{\partial {\Gamma^1}_{11}}{\partial r} = \frac{\partial}{\partial r}\frac{V'}{2V}$$

D_{11} is given by

$$D_{11} = \frac{\partial}{\partial r}\left(\frac{U'}{2U} + \frac{V'}{2V} + \frac{2}{r}\right)$$

As $Ric_{11} = A_{11} - B_{11} + C_{11} - D_{11}$,

$$\begin{aligned} Ric_{11} &= \frac{U'V'}{4UV} + \frac{V'^2}{4V^2} + \frac{V'}{rV} - \frac{U'^2}{4U^2} - \frac{V'^2}{4V^2} - \frac{2}{r^2} \\ &+ \frac{\partial}{\partial r}\frac{V'}{2V} - \frac{\partial}{\partial r}\frac{U'}{2U} - \frac{\partial}{\partial r}\frac{V'}{2V} - \frac{\partial}{\partial r}\frac{2}{r} \end{aligned}$$

Some terms get canceled:

$$\begin{aligned} Ric_{11} &= \frac{U'V'}{4UV} + \frac{V'}{rV} - \frac{U'^2}{4U^2} - \frac{\partial}{\partial r}\frac{U'}{2U} \\ &= \frac{U'V'}{4UV} + \frac{V'}{rV} - \frac{U'^2}{4U^2} - \frac{U''}{2U} + \frac{U'^2}{2U^2} \\ &= \frac{U'V'}{4UV} + \frac{V'}{rV} - \frac{U''}{2U} + \frac{U'^2}{4U^2} \end{aligned} \tag{8.32}$$

The next component is Ric_{22}:

$$Ric_{22} = A_{22} - B_{22} + C_{22} - D_{22} = \Gamma^k{}_{lek}\Gamma^l{}_{22} - \Gamma^k{}_{l2}\Gamma^l{}_{2k} + \frac{\partial \Gamma^k{}_{22}}{\partial u^k} - \frac{\partial \Gamma^k{}_{2k}}{\partial \theta}$$

The only non-zero term in the first term A_{22} is for $l = 1$:

$$\begin{aligned} A_{22} &= \Gamma^k{}_{1k}\Gamma^1{}_{22} \\ &= \left(\frac{U'}{2U} + \frac{V'}{2V} + \frac{2}{r}\right)\left(-\frac{r}{V}\right) \\ &= -\frac{rU'}{2UV} - \frac{rV'}{2V^2} - \frac{2}{V} \end{aligned}$$

For B_{22}, expand the terms for $l = 1, 2, 3$:

$$B_{22} = \Gamma^k{}_{12}\Gamma^1{}_{2k} + \Gamma^k{}_{22}\Gamma^2{}_{2k} + \Gamma^k{}_{32}\Gamma^3{}_{2k}$$

Looking up the non-zero Christoffel symbols terms for the lower indices $12, 22, 32$,

$$\begin{aligned} B_{22} &= \Gamma^2{}_{12}\Gamma^1{}_{22} + \Gamma^1{}_{22}\Gamma^2{}_{21} + \Gamma^3{}_{32}\Gamma^3{}_{23} \\ &= 2\Gamma^2{}_{12}\Gamma^1{}_{22} + \Gamma^3{}_{32}\Gamma^3{}_{23} \\ &= -2\frac{1}{r}\frac{r}{V} + \cot^2\theta \\ &= -\frac{2}{V} + \cot^2\theta \end{aligned}$$

For C_{22}, the only non-zero term is when $l = 1$:

$$\begin{aligned} C_{22} &= \frac{\partial \Gamma^1{}_{22}}{\partial r} \\ &= \frac{\partial}{\partial r}\left(-\frac{r}{V}\right) \\ &= -\frac{1}{V} + \frac{rV'}{V^2} \end{aligned}$$

For D_{22}, using Eq. (8.31),

$$D_{22} = \frac{\partial \Gamma^k{}_{2k}}{\partial \theta} = \frac{\partial \cot\theta}{\partial \theta} = -csc^2\theta$$

As $Ric_{22} = A_{22} - B_{22} + C_{22} - D_{22}$ and using $\csc^2\theta - \cot^2\theta = 1$,

$$\begin{aligned} Ric_{22} &= -\frac{rU'}{2UV} - \frac{rV'}{2V^2} - \frac{2}{V} + \frac{2}{V} - \cot^2\theta - \frac{1}{V} + \frac{rV'}{V^2} + csc^2\theta \\ &= -\frac{rU'}{2UV} - \frac{1}{V} + \frac{rV'}{2V^2} + 1 \end{aligned} \tag{8.33}$$

Lastly, we evaluate Ric_{33}:

$$Ric_{33} = A_{33} - B_{33} + C_{33} - D_{33} = \Gamma^k{}_{lek}\Gamma^l{}_{33} - \Gamma^k{}_{l3}\Gamma^l{}_{3k} + \frac{\partial \Gamma^k{}_{33}}{\partial u^k} - \frac{\partial \Gamma^k{}_{3k}}{\partial \phi}$$

For A_{33}, the non-zero terms occur when $l = 1, 2$. Expand this term for $l = 1, 2$:

$$\begin{aligned} A_{33} &= \Gamma^k{}_{1k}\Gamma^1{}_{33} + \Gamma^k{}_{2k}\Gamma^2{}_{33} \\ &= \left(\frac{U'}{2U} + \frac{V'}{2V} + \frac{2}{r}\right)\left(-\frac{r\sin^2\theta}{V}\right) + \cot\theta(-\sin\theta\cos\theta) \\ &= -\frac{rU'\sin^2\theta}{2UV} - \frac{rV'\sin^2\theta}{2V^2} - \frac{2\sin^2\theta}{V} - \cos^2\theta \end{aligned}$$

For B_{33}, expand the term for $l = 1, 2, 3$:

$$B_{33} = \Gamma^k{}_{13}\Gamma^1{}_{3k} + \Gamma^k{}_{23}\Gamma^2{}_{3k} + \Gamma^k{}_{33}\Gamma^3{}_{3k}$$

The first term with the lower indices 13 has $k = 3$ as a non-zero term. The second term with the lower indices 23 has $k = 3$ as a non-zero term. The last term has $k = 1, 2$ as a non-zero term:

$$\begin{aligned} B_{33} &= \Gamma^3{}_{13}\Gamma^1{}_{33} + \Gamma^3{}_{23}\Gamma^2{}_{33} + \Gamma^1{}_{33}\Gamma^3{}_{31} + \Gamma^2{}_{33}\Gamma^3{}_{32} \\ &= -\frac{1}{r}\frac{r\sin^2\theta}{V} - \cot\theta\sin\theta\cos\theta - \frac{r\sin^2\theta}{V}\frac{1}{r} - \sin\theta\cos\theta\cot\theta \\ &= -2\frac{\sin^2\theta}{V} - 2\cos^2\theta \end{aligned} \tag{8.34}$$

C_{33} has a non-zero term for $k = 1, 2$:

$$\begin{aligned} C_{33} &= \frac{\partial \Gamma^1{}_{33}}{\partial r} + \frac{\partial \Gamma^2{}_{33}}{\partial \theta} \\ &= -\frac{\partial}{\partial r}\left(\frac{r\sin^2\theta}{V}\right) - \frac{\partial}{\partial \theta}(\sin\theta\cos\theta) \\ &= -\frac{\sin^2\theta}{V} + \frac{rV'\sin^2\theta}{V^2} - \cos^2\theta + \sin^2\theta \end{aligned}$$

$D_{33} = 0$ because the partial differential of Christoffel symbols with respect to ϕ is zero. So, $Ric_{33} = A_{33} - B_{33} + C_{33} - D_{33}$:

$$\begin{aligned} Ric_{33} = &-\frac{rU'\sin^2\theta}{2UV} - \frac{rV'\sin^2\theta}{2V^2} - \frac{2\sin^2\theta}{V} - \cos^2\theta \\ &+2\frac{\sin^2\theta}{V} + 2\cos^2\theta - \frac{\sin^2\theta}{V} + \frac{rV'\sin^2\theta}{V^2} - \cos^2\theta + \sin^2\theta \\ = &-\frac{rU'\sin^2\theta}{2UV} + \frac{rV'\sin^2\theta}{2V^2} - \frac{\sin^2\theta}{V} + \sin^2\theta \end{aligned} \tag{8.35}$$

In terms of the unknown functions $U(r), V(r)$, and r, θ, the Ricci tensor evaluate to the following:

$$Ric = \begin{pmatrix} \frac{U''}{2V} + \frac{U'}{rV} - \frac{U'V'}{4V^2} - \frac{U'^2}{4UV} & 0 & 0 & 0 \\ 0 & -\frac{U''}{2U} + \frac{U'V'}{4UV} + \frac{U'^2}{4U^2} + \frac{V'}{rV} & 0 & 0 \\ 0 & 0 & -\frac{rU'}{2UV} - \frac{1}{V} + \frac{rV'}{2V^2} + 1 & 0 \\ 0 & 0 & 0 & \left(-\frac{rU'}{2UV} + \frac{rV'}{2V^2} - \frac{1}{V} + 1\right)\sin^2\theta \end{pmatrix} \tag{8.29}$$

The off-diagonal elements are evaluated to be zero. We leave it as an exercise for the reader to verify this.

Einstein field equations

Using $Ric_{ij} = 0$, we solve the differential equations for $U(r)$ and $V(r)$. Using Ric_{00} in Eq. (8.31), Ric_{11} in Eq. (8.32), Ric_{22} in Eq. (8.33), and Ric_{33} in Eq. (8.35),

$$Ric_{00} = \frac{U''}{2V} + \frac{U'}{rV} - \frac{U'^2}{4UV} - \frac{U'V'}{4V^2} = 0$$

$$Ric_{11} = \frac{U'V'}{4UV} + \frac{V'}{rV} - \frac{U''}{2U} + \frac{U'^2}{4U^2} = 0$$

$$Ric_{22} = -\frac{rU'}{2UV} - \frac{1}{V} + \frac{rV'}{2V^2} + 1 = 0$$

$$Ric_{33} = -\frac{rU'\sin^2\theta}{2UV} + \frac{rV'\sin^2\theta}{2V^2} - \frac{\sin^2\theta}{V} + \sin^2\theta = 0$$

Using $VRic_{00} + URic_{11} = 0$, we arrive at

$$\begin{aligned} \frac{U'}{r} + \frac{V'U}{rV} &= 0 \\ U'V + V'U &= 0 \\ \frac{\partial UV}{\partial r} &= 0 \\ UV &= c \text{ (const)} \end{aligned}$$

In the limit $r \to \infty$, we have the Minkowski metric in polar coordinates, which is

$$g = \begin{pmatrix} 1 & 0 & 0 & 0 \\ 0 & -1 & 0 & 0 \\ 0 & 0 & -r^2 & 0 \\ 0 & 0 & 0 & -r^2\sin^2\theta \end{pmatrix}$$

with $U = V = 1$ and $UV = 1$. Therefore, $c = 1$. We arrive at the first important conclusion:

$$U = 1/V \tag{8.36}$$

Also,

$$\frac{V'}{V^2} = -U'$$

Substitute Eq. (8.36) into $Ric_{22} = 0$:

$$\begin{aligned} Ric_{22} &= -\frac{rU'}{2UV} - \frac{1}{V} + \frac{rV'}{2V^2} + 1 = 0 \\ & -\frac{rU'}{2} - U - \frac{rU'}{2} + 1 = 0 \\ & -rU' - U + 1 = 0 \\ & rU' = 1 - U \end{aligned}$$

The solution for this differential equation is

$$U(r) = 1 - \frac{r_s}{r}$$

The Scharwzchild metric is then

$$g = \begin{pmatrix} \left(1-\frac{r_s}{r}\right) & 0 & 0 & 0 \\ 0 & -\left(1-\frac{r_s}{r}\right)^{-1} & 0 & 0 \\ 0 & 0 & -r^2 & 0 \\ 0 & 0 & 0 & -r^2\sin^2\theta \end{pmatrix}$$

When $r = r_s$, the time component $g_{00} = 0$. What does that mean? It means the time measure becomes zero. Time stops. The space component in the radial direction $g_{11} \to \infty$, which means distance goes to ∞. We call r_s the Schwarzschild radius. Next, we use Ric_{33} to check our solutions:

$$\begin{aligned} rU' &= 1 - U \\ U &= 1/V \\ U' &= -V'/V^2 \end{aligned}$$

$$\begin{aligned} Ric_{33} &= \left(-\frac{rU'}{2UV} + \frac{rV'}{2V^2} - \frac{1}{V} + 1\right)\sin^2\theta \\ &= \left(-\frac{rU'}{2} - \frac{rU'}{2} - U + 1\right)\sin^2\theta \\ &= (-rU' - U + 1)\sin^2\theta \\ &= (U - 1 - U + 1)\sin^2\theta = 0 \end{aligned}$$

Schwarzschild radius

We derive an approximation for the Schwarzschild radius. Consider a test mass far away from the black hole, $r \gg r_s$. It is initially at rest in space, i.e. $\dot{r} = \dot{\theta} = \dot{\phi} = 0$, and accelerating toward the black hole due to gravity. Then, the acceleration using Newton's gravitational law is

$$\frac{d^2r}{dt^2} = -\frac{GM}{r^2}$$

Consider this motion as a geodesic motion in curved space-time. Then, we have the geodesic equation on space-time coordinates (ct, r, θ, ϕ), and write the geodesic equation in terms of time coordinate:

$$\frac{d^2r}{dt^2} + \Gamma^1{}_{ij}\frac{du^i}{dt}\frac{du^j}{dt} = \Gamma^0{}_{ij}\frac{du^i}{dt}\frac{du^j}{dt}\frac{dr}{dt}$$

We have not derived this equation. We have $d\theta/dt = d\phi/dt = 0$, as the particle moves radially only. Also, since the particle is far away and we assume the non-relativistic case, $dr/dt \ll c$. We write

$$\frac{d^2r}{dt^2} + \Gamma^1{}_{00}c^2 \approx 0$$

Comparing this equation with Eq. (6):

$$\Gamma^1{}_{00}c^2 = \frac{U'c^2}{2V} \approx \frac{GM}{r^2}$$

Substitute $U = 1 - r_s/r$, $U = 1/V$ and make the approximation $r_s/r \ll 1$:

$$U' = \frac{r_s}{r^2}$$

$$\begin{aligned} \frac{r_s}{2r^2}\left(1 - \frac{r_s}{r}\right)c^2 &\approx \frac{GM}{r^2} \\ r_s &\approx \frac{2GM}{c^2} \end{aligned}$$

Different sources derive r_s in slightly different ways, all of which use some approximations. r_s is also called the event horizon of a black hole.

8.6 Summary

Components of the Riemann curvature tensor

The Riemann curvature tensor is, as given in Eq. (8.6),

$$R^k{}_{imj} = \Gamma^k{}_{lm}\Gamma^l{}_{ij} - \Gamma^k{}_{lj}\Gamma^l{}_{im} + \frac{\partial\Gamma^k{}_{ij}}{\partial u^m} - \frac{\partial\Gamma^k{}_{im}}{\partial u^j} \tag{8.6}$$

or, as in Eq. (8.11),

$$R_{simj} = \Gamma^k{}_{im}\Gamma_{ksj} - \Gamma^k{}_{ij}\Gamma_{ksm} + \frac{\partial\Gamma_{sij}}{\partial u^m} - \frac{\partial\Gamma_{sim}}{\partial u^j} \tag{8.11}$$

Written in terms of the tensor product of covectors and vectors,

$$R = R^k{}_{imj}du^m du^j du^i \partial_{u^k} \tag{8.7}$$

Written in the form of covariant derivatives,

$$R(X,Y) = [D_X, D_Y] - D_{[X,Y]} \tag{8.10}$$

Properties of the Riemann curvature tensor

Lemma 8.3.1 *Properties of ${R^k}_{imj}$:*

1. *Antisymmetric property:*

$$\begin{aligned} {R^k}_{imj} &= -{R^k}_{ijm} \\ R_{kijm} &= -R_{kimj} \\ R_{kijm} &= -R_{ikjm} \end{aligned} \tag{8.12}$$

2. *Antisymmetric in the final three indices:*

$${R^k}_{imj} + {R^k}_{jim} + {R^k}_{mji} = 0 \tag{8.13}$$

3. *Block symmetry:*

$$R_{kijm} = R_{jmki} \tag{8.14}$$

4. *From the antisymmetric property:*

$$\begin{aligned} {R^k}_{ijj} &= 0 \\ R_{kijj} &= 0 \end{aligned} \tag{8.15}$$

Jacobi equation

As given in Eq. (8.20),

$$D_{\dot{\gamma}} D_{\dot{\gamma}} J - R(\dot{\gamma}, J)\dot{\gamma} = 0 \tag{8.20}$$

Sectional curvature, Ricci tensor and scalar curvature

Sectional curvature:

$$K(X, Y) = \frac{R(X, Y)X \cdot Y}{(X \cdot X)(Y \cdot Y) - (X \cdot Y)^2} \tag{8.21}$$

The Ricci tensor:

$$Ric(\partial_{u^d}, \partial_{u^d}) = \frac{1}{d-1} \sum_{i}^{d-1} R(\partial_{u^d}, \partial_{u^i})\partial_{u^d} \cdot \partial_{u^i} \in \mathbb{R} \tag{8.22}$$

$$Ric_{lm} = {R^i}_{lim} \tag{8.23}$$

Scalar curvature:

$$S = g^{ij} R_{ij} \tag{8.26}$$

Chapter 9

Putting It All Together

In this chapter, we summarize all the topics that are covered in this book as well as highlight connections between the different topics. This chapter gives a big picture of the basics of differential geometry. It can help the reader connect the dots.

9.1 Equations, Definitions, Lemmas and Theorems

In this section, we list all the definitions, lemmas and theorems in this book for the reader's quick reference. Readers should note that the definitions, lemmas and theorems repeated here are identical to those in previous chapters; hence, we have enumerated them identically to those in the previous chapters, where they first appear.

Definition 2.4.1 Given n types of elements $e_i, i = 1, \ldots, n$, define the binary operations $\cdot$, $+$ and scalar multiplication such that

$$\alpha e_i + \beta e_i = (\alpha + \beta) e_i, \qquad \alpha, \beta \in \mathbb{R}$$
$$e_i \cdot e_j = g_{ij} \qquad g_{ij} \in \mathbb{R}$$

g_{ij} is called the metric tensor.

Lemma 2.4.1 *Given a coordinate transformation* $\hat{e}_i = J^k{}_i e_k$*, the components of* e_k *transform as* $\hat{x}^i = {J^{-1}}^i{}_k x^k$ *with* $J^k{}_i {J^{-1}}^i{}_j = \delta^k{}_j$.

Lemma 2.4.2 *The metric tensor transform as*

$$\hat{g} = J^T g J$$

Lemma 2.4.3 *The vector length is invariant with respect to the coordinate transformation:*

$$\hat{x} \cdot \hat{x} = x \cdot x$$

Definition 3.5.1 Given a differentiable mapping $\hat{x} = \hat{x}(x)$, the Jacobian is defined as

$$J^j{}_i = \frac{\partial x^j}{\partial \hat{x}^i}$$

Definition 3.5.2 Polar coordinates defined on $\mathbb{R}^2$ are given by

$$\begin{aligned} q^1 &= r\cos\theta \\ q^2 &= r\sin\theta \end{aligned}$$

Definition 3.5.3 The arc length of a path parameterized by $t \in \mathbb{R}$ is given by

$$l = \int_0^1 \sqrt{\frac{dx}{dt} \cdot \frac{dx}{dt}}\, dt$$

where the inner product involves using the metric tensor.

Definition 4.2.1 A chart φ is a differentiable and invertible mapping from $U \subset \mathbb{R}^d \mapsto \mathbb{R}^n$, such that $u \in U$ and $\varphi(u) \in M$.

Definition 4.3.1 A tangent space $T_x M = \mathbb{R}^d$ at the point $x \in M$ is spanned by basis vectors:

$$\hat{e}_i = \frac{\partial \varphi(u)}{\partial u^i}, \qquad i = 1, \ldots, d$$

Definition 4.3.2 (differentiable tangent vector field) The tangent vectors at each point of M are given by

$$X = X^i(u)\hat{e}_i = X^i(u)\frac{\partial \varphi}{\partial u^i}$$

Its components, $X^i : U \mapsto \mathbb{R}$, are differentiable functions.

Definition 4.3.3 At every point on a path $\gamma(t)$ in M, there is a corresponding tangent vector $X(t)$:

$$X(t) = X^i\hat{e}_i = \dot{\gamma}(t) = \dot{\gamma}^i\hat{e}_i$$

Lemma 4.6.1 *A path which is a geodesic is given by the solution of the geodesic differential equation*

$$\frac{d^2u^k}{dt^2} + \Gamma^k{}_{ij}\frac{du^j}{dt}\frac{du^i}{dt} = 0$$

Definition 4.7.1 The induced metric tensor of $M \subset \mathbb{R}^n$ is given by

$$g_{ij} = \hat{e}_i \cdot \hat{e}_j$$

where the dot product is given by the Euclidean dot product.

Definition 4.7.2 A geodesic is a path with the shortest length defined in Definition (3.5.3) with appropriate initial or boundary conditions,

Definition 5.7.1 The covariant derivative of a vector Y in the direction of X is given by

$$D_X Y = X^j \left(\frac{\partial Y^k}{\partial u^j} + Y^i \Gamma^k{}_{ij} \right) \hat{e}_k \in T_x M \tag{5.60}$$

Definition 5.7.2 The covariant derivative on a path $\gamma(t)$ is given by

$$D_{\dot{\gamma}} X = \left(D_{\hat{e}_j} X\right) \frac{du^j}{dt} = \left(D_{\hat{e}_j} X\right) \dot{\gamma}^j \tag{5.61}$$

Lemma 5.3.1 *Let $D_Y X$ be the covariant derivative of X in the direction $Y \in T_x M$. Let $Z \in T_x M$ and $X \in T_x M$. Then, the following identities hold:*

1. *Linear in the direction of derivative: $f, g : M \mapsto \mathbb{R}$.*

$$D_{fY_1 + gZ_2} X = f D_Y X + g D_Z X \tag{5.4}$$

2. *Linear in input: $a_1, a_2 \in \mathbb{R}$.*

$$D_Z(a_1 X + a_2 Y) = a_1 D_Z X + a_2 D_Z Y \tag{5.5}$$

3. *Product rule, $f : M \mapsto \mathbb{R}$:*

$$D_Y(fX) = f D_Y X + X D_Y f \tag{5.6}$$

4. *Metric compatibility:*

$$D_Z(X \cdot Y) = (D_Z X) \cdot Y + X \cdot (D_Z Y) \tag{5.7}$$

5. *Torsion-free property:*

$$D_X Y - D_Y X = [X, Y] = \left(X^i \frac{\partial Y^j}{\partial u^i} - Y^i \frac{\partial X^j}{\partial u^i} \right) \hat{e}_j \tag{5.8}$$

Definition 5.7.3 Given a path $\gamma(t)$, a vector field $X(t)$ along $\gamma(t)$ is being parallel transported if

$$D_{\dot{\gamma}}X(t) = 0 \qquad \forall t \in [0,1] \tag{5.62}$$

Lemma 5.6.1 *These properties of parallel transport hold:*

1. *Given two vector fields $X(t)$ and $Y(t)$, in which they are parallel transported along a path $\gamma(t)$ in M, then their inner product is constant:*

$$X(0) \cdot Y(0) = X(t) \cdot Y(t) \qquad \forall t \tag{5.49}$$

2. *Given a path $\gamma(t)$, if*

$$D_{\dot{\gamma}(t)}\dot{\gamma}(t) = 0 \qquad \forall t \tag{5.50}$$

then the path $\gamma(t)$ traces out a geodesic.

Definition 5.7.4 Given a geodesic $\gamma(t)$ with $\gamma(0) = x$, $\dot{\gamma}(0) = X$, the exponential map is a map $\exp : T_xM \mapsto M$ given by

$$\exp(X) = \gamma(1) \tag{5.63}$$

Lemma 5.7.1 *If $\gamma(t)$ be a geodesic such that*

$$D_{\dot{\gamma}}\dot{\gamma} = 0 \tag{5.64}$$

and reparameterizing $\tau = \alpha t$, then $\gamma(\tau)$ is a geodesic with $\dot{\gamma}(\tau) = \alpha\dot{\gamma}(t)$.

Definition 6.2.1 (tangent vectors) Given a manifold U, with $u \in U$, define the basis tangent vector as

$$\hat{e}_i = \frac{\partial}{\partial u^i} = \partial_{u^i}$$

Tangent vectors are given by $X = X^i\partial_{u^i}$, which is a linear mapping

$$X : C^\infty(U) \mapsto \mathbb{R}$$

C^∞ is the set of differentiable functions that maps U to $\mathbb{R}$.

Definition 6.2.2 Tangent vectors can be defined as the derivative of a path on the manifold:

$$\frac{d\gamma(t)}{dt} \equiv \frac{du^i}{dt}\frac{\partial}{\partial u^i} = X^i\partial_{u^i} \tag{6.2}$$

Definition 6.3.1 (Covectors) Given a scalar function $f : U \mapsto \mathbb{R}$, define its differential df as a map $df : T_u U \mapsto \mathbb{R}$ as

$$df(X) = X(f) \qquad \forall X \in T_u U$$

Lemma 6.4.1 *The covariant derivative of the metric tensor is zero:*

$$D_{\partial_{u^i}} g = 0$$

Definition 6.5.1 (pushforward transformation) Given two manifolds U and $\hat{U}$ and a differentiable map between them φ, and given a vector $X = X^i \partial_{u^i} \in T_u U$, the pushforward of this vector onto $\hat{U}$ via φ is given by the contravariant transformation of its components with X applied to the scalar functions $\hat{f} \circ \varphi$, $\hat{f} : \hat{U} \mapsto \mathbb{R}$. Let the pushforward vector be $\hat{X} = \hat{X}^i \partial_{u^i}$. Then,

$$\hat{X}^j(\varphi(u)) = (J^{-1})_i^{\ j} X^i(u)$$

$$X(f) = X(\hat{f} \circ \varphi) = \hat{X}(\hat{f})$$

Lemma 6.5.1 *Given maps between manifolds $\varphi : U \mapsto \hat{U}$ and $\hat{\varphi} : \hat{U} \mapsto \tilde{U}$. The pushforward between these maps is given by*

$$(\hat{\varphi} \circ \varphi)_* = \hat{\varphi}_* \circ \varphi_*$$

Definition 6.5.2 (pullback of covariant tensors) Let $\hat{\tau}_{\hat{u}}$ be a tensor acting on vectors $X, Y, Z, \ldots$ in the vector space $T_{\hat{u}}\hat{U}$ in $\hat{U}$:

$$\hat{\tau}_u : T_{\hat{u}}\hat{U} \times \cdots \times T_{\hat{u}}\hat{U} \mapsto \mathbb{R}$$

Given a differentiable map $\varphi : U \mapsto \hat{U}$, with $\hat{u} = \varphi(u)$, the pullback of $\hat{\tau}_{\hat{u}}$ under φ is given by

$$(\varphi^* \hat{\tau})_u(X, Y, Z, \ldots) = \hat{\tau}_{\hat{u}}(\varphi_* X, \varphi_* Y, \varphi_* Z, \ldots)$$

Theorem 7.1.1 *Let X be a smooth vector field on U. Then, the flow path generated by X starting from u_0 and the pushforward of X under $\sigma_{X,t}$ satisfy,*[1]

$$\sigma_{X,t}(u_0)_* X_{u_0} = X_{\sigma_{X,t}(u_0)} \tag{7.3}$$

Theorem 7.1.2 *Let X be a vector field in U with flow $\sigma_{X,t}$ and $\hat{X}$ be a vector field in $\hat{U}$ with flow $\sigma_{\hat{X},t}$. Let $\varphi : U \mapsto \hat{U}$. Then, φ_* relates to the flows by*[2]

$$\sigma_{\hat{X},t} \circ \varphi = \varphi \circ \sigma_{X,t} \iff \varphi_* X = \hat{X}. \tag{7.4}$$

[1]See John Lee, *Introduction to Smooth Manifolds*, p. 442 [8].
[2]See John Lee, *Introduction to Smooth Manifolds*, p. 468 [8].

The Lie derivative is defined as follows:

$$\mathcal{L}_X Y = \lim_{\epsilon \to 0} \frac{(\sigma_{X,-\epsilon})_* Y(u_\epsilon) - Y(u_0)}{\epsilon} \tag{7.12}$$

Definition 7.6.4 (Lie derivative) The Lie derivative of a vector field Y in the direction of another vector field is X,

$$\mathcal{L}_X Y = \left(X^j \frac{\partial Y^i}{\partial u^j} - Y^j \frac{\partial X^i}{\partial u^j} \right) \partial_{u^i}$$

The Lie derivative of a covector field ω is

$$\mathcal{L}_X \omega = \left(X^i \frac{\partial \omega_j}{\partial u^i} + \omega_i \frac{\partial X^i}{\partial u^j} \right) du^j$$

Lemma 7.6.1 *The Lie derivative is equal to the Lie bracket:*

$$\mathcal{L}_X Y = [X, Y]$$

Lemma 7.5.1 (Properties of Lie derivatives) *Some properties of Lie derivatives and the Lie bracket are stated as follows:*

1. *Antisymmetric:*

$$\begin{aligned} \mathcal{L}_X Y &= -\mathcal{L}_Y X \\ [X, Y] &= -[X, Y] \end{aligned} \tag{7.31}$$

2. *Bilinear:*

$$\begin{aligned} [X, Y + Z] &= [X, Y] + [X, Z] \\ [X + Y, Z] &= [X, Z] + [Y, Z] \end{aligned}$$

3. *The Jacobi identity, i.e. the sum of cyclic permutations is zero:*

$$[X, [Y, Z]] + [Z, [X, Y]] + [Y, [Z, X]] = 0 \tag{7.32}$$

4. *Product rule:*

$$[X, fY] = X(f)Y + f[X, Y] \tag{7.33}$$

5. *Commutator of Lie derivatives:*

$$[\mathcal{L}_X, \mathcal{L}_Y] = \mathcal{L}_{[X,Y]}$$

Lemma 7.5.2 *The Lie derivative can be written in the form of covariant derivatives:*

$$\mathcal{L}_X Y = D_X Y - D_Y X = [X, Y] \qquad \text{(torsion-free property)}$$

Definition 7.5.1 Let X and Y be two vector fields in U. Let their respective flows be $\sigma_{X,t}$ and $\sigma_{Y,s}$. The flows commute if for all $u \in U$,

$$\sigma_{X,t} \circ \sigma_{Y,s}(u) = \sigma_{Y,s} \circ \sigma_{X,t}(u)$$

Theorem 7.5.1 (commuting flows) *Let X and Y be two vector fields in U with the respective flows as $\sigma_{X,t}$ and $\sigma_{Y,s}$. The flows commute iff $[X, Y] = 0$.*

9.1.1 The Riemann curvature tensor

$$R^k{}_{imj} = \Gamma^k{}_{lm}\Gamma^l{}_{ij} - \Gamma^k{}_{lj}\Gamma^l{}_{im} + \frac{\partial \Gamma^k{}_{ij}}{\partial u^m} - \frac{\partial \Gamma^k{}_{im}}{\partial u^j} \tag{8.6}$$

$$R = R^k{}_{imj} du^m du^j du^i \partial_{u^k} \tag{8.7}$$

$$R(X, Y) = [D_X, D_Y] - D_{[X,Y]} \tag{8.10}$$

$$R_{simj} = \Gamma^k{}_{im}\Gamma_{ksj} - \Gamma^k{}_{ij}\Gamma_{ksm} + \frac{\partial \Gamma_{sij}}{\partial u^m} - \frac{\partial \Gamma_{sim}}{\partial u^j} \tag{8.11}$$

Lemma 8.3.1 *Properties of $R^k{}_{imj}$:*

1. *Antisymmetric property:*

$$\begin{aligned} R^k{}_{imj} &= -R^k{}_{ijm} \\ R_{kijm} &= -R_{kimj} \\ R_{kijm} &= -R_{ikjm} \end{aligned} \tag{8.12}$$

2. *Antisymmetric in the final three indices:*

$$R^k{}_{imj} + R^k{}_{jim} + R^k{}_{mji} = 0 \tag{8.13}$$

3. *Block symmetry:*

$$R_{kijm} = R_{jmki} \tag{8.14}$$

4. *From the antisymmetric property:*

$$\begin{aligned} R^k{}_{ijj} &= 0 \\ R_{kijj} &= 0 \end{aligned} \tag{8.15}$$

Table 8.1. Non-zero Christoffel symbols.

Upper index	Lower indices	value
0	01,10	${\Gamma^0}_{10} = {\Gamma^0}_{01} = \frac{U'}{2U}$
1	00	${\Gamma^1}_{00} = \frac{U'}{2V}$
	11	${\Gamma^1}_{11} = \frac{V'}{2V}$
	22	${\Gamma^1}_{22} = -\frac{r}{V}$
	33	${\Gamma^1}_{33} = -\frac{r\sin^2\theta}{V}$
2	12,21	${\Gamma^2}_{21} = {\Gamma^2}_{12} = \frac{1}{r}$
	33	${\Gamma^2}_{33} = -\sin\theta\cos\theta$
3	13,31	${\Gamma^3}_{13} = {\Gamma^3}_{31} = \frac{1}{r}$
	23,32	${\Gamma^3}_{23} = {\Gamma^3}_{32} = \cot\theta$

9.1.2 Jacobi fields

$$D_{\dot\gamma} D_{\dot\gamma} J - R(\dot\gamma, J)\dot\gamma = 0 \tag{8.20}$$

9.1.3 Sectional curvature, Ricci tensor and Ricci scalar

$$K(X,Y) = \frac{R(X,Y)X \cdot Y}{(X \cdot X)(Y \cdot Y) - (X \cdot Y)^2} \tag{8.21}$$

$$Ric(\partial_{u^d}, \partial_{u^d}) = \frac{1}{d-1} \sum_i^{d-1} R(\partial_{u^d}, \partial_{u^i})\partial_{u^d} \cdot \partial_{u^i} \in \mathbb{R} \tag{8.22}$$

$$Ric_{lm} = {R^i}_{lim} \tag{8.23}$$

$$S = g^{ij} R_{ij} \tag{8.26}$$

9.1.4 Einstein field equations and the Schwarzschild solution

$$Ric_{ij} - \frac{1}{2} S g_{ij} + \Lambda g_{ij} = \frac{8\pi G}{c^4} T_{ij} \tag{8.28}$$

$$Ric = \begin{pmatrix} \frac{U''}{2V} + \frac{U'}{rV} - \frac{U'V'}{4V^2} - \frac{U'^2}{4UV} & 0 & 0 & 0 \\ 0 & -\frac{U''}{2U} + \frac{U'V'}{4UV} + \frac{U'^2}{4U^2} + \frac{V'}{rV} & 0 & 0 \\ 0 & 0 & -\frac{rU'}{2UV} - \frac{1}{V} + \frac{rV'}{2V^2} + 1 & 0 \\ 0 & 0 & 0 & \left(-\frac{rU'}{2UV} + \frac{rV'}{2V^2} - \frac{1}{V} + 1\right) \sin^2\theta \end{pmatrix} \quad (8.29)$$

	Concept	Comments
(I)	Chart $\varphi : U \mapsto U'$	Mapping between surfaces
(II)	Metric Tensor g_{ij}	Measure of lengths; a fundamental quantity which is not derived from other quantities
(III)	Christoffel symbols $\Gamma^k{}_{ij} = \frac{1}{2} g^{ks} \left(\frac{\partial g_{si}}{\partial u^j} + \frac{\partial g_{sj}}{\partial u^i} - \frac{\partial g_{ij}}{\partial u^s} \right)$	Derived from the metric tensor (II)
(IV)	Tangent vectors $X = X^i \partial_{u^i}, X : \mathbb{F} \mapsto \mathbb{R}$	Differential with respect to coordinate representation φ. Maps functions ($f : U \mapsto \mathbb{R}, f \in \mathbb{F}$) onto the real number
(V)	Covectors $\omega = \omega_i du^i$ $\omega : X \mapsto \mathbb{R}$	Dual of the tangent vector Maps tangent vectors onto the real number
(VI)	Geodesics $\frac{d^2u^k}{dt^2} + \Gamma^k{}_{ij} \frac{du^i}{dt} \frac{du^j}{dt} = 0$	Shortest distance between two points on a surface. Uses the Christoffel symbols (III)
(VII)	Covariant derivative $D_X Y = X^j \left(\frac{\partial Y^k}{\partial u^j} + Y^i \Gamma^k{}_{ij} \right) \partial_{u^k}$	Differential in the direction of another tangent vector. Covariant derivative can be computed for vectors, covectors and tensors in general. Uses the Christoffel symbols (III), tangent vectors (IV) and covectors (V).
(VIII)	Covariant derivative of metric tensor $D_X g = 0$	This quantity is always zero.

	Concept	Comments
(IX)	Parallel transport $D_{\dot{\gamma}}X = 0$	To get "parallel vectors" on curved surfaces. Uses the covariant derivative (VII).
(X)	Exponential map $\mathrm{Exp}(X) = \gamma(1)$	Map a tangent vector onto a point on the surface using the geodesic equation (VI).
(XI)	Pushforward $(\varphi_* X)(\hat{f}) = \hat{X}(\hat{f}) = X(f \circ \varphi)$	Maps a vector (IV) from one surface onto another.
(XII)	Pullback $(\varphi^* \hat{\omega})_u(X, \ldots) =$ $\hat{\omega}_{\varphi(u)}(\varphi_* X, \ldots)$	Maps a covector (or tensor) (V) from one surface onto another.
(XIII)	Flows $\sigma_{X,t}(u)$	Lines traced out by a tangent to vector field.
(XIV)	Pushforward, map and flows $(\sigma_{X,t})_* X = X_{\sigma_{X,t}}$ $\sigma_{\hat{X},t} \circ \varphi = \varphi \circ \sigma_{X,t} \Leftrightarrow \varphi_* X =$ $\hat{X}$	Relationship between flows (XIII) and pushforward (XI).
(XV)	Lie derivative $\mathcal{L}_X Y = [X, Y]$	Differentiate along a flow direction. Uses pushforward (XI) and pullback (XII).
(XVI)	Commuting flows $\sigma_{Y,s} \circ \sigma_{X,t} = \sigma_{X,t} \circ \sigma_{Y,s} \Leftrightarrow$ $[X, Y] = 0$	An important result linking the Lie derivative (XV) and flows (XIII).
(XVII)	Riemann curvature tensor $R = [D_X, D_Y] - D_{[X,Y]}$	Measure of curvature of a surface. Uses parallel transport (IX), Covariant derivative (VIII), Lie derivative (XV).
(XVIII)	Jacobi equations $D_{\dot{\gamma}} D_{\dot{\gamma}} J - R(\dot{\gamma}, J)\dot{\gamma} = 0$	Measure of geodesic deviation.
(XIX)	Ricci tensor $R_{ij} = R^k{}_{ikj}$	Describes how a volume of space-time changes when this volume move along a geodesic.

	Concept	Comments
(XX)	Ricci scalar $S = R_{ij}g^{ij}$	A kind of average curvature measured in all directions.
(XI)	Einstein field equations $R_{ij} - \frac{1}{2}Sg_{ij} + \Lambda g_{ij} = \frac{8\pi G}{c^4}T_{ij}$	Gives the relationship between the metric of curved space-time, energy and mass. Uses the Ricci tensor (XIX), Ricci scalar (XX), metric tensor (II).

Bibliography

[1] C. Blinn. Schwarzschild solution to Einstein's general relativity. https://sites.math.washington.edu/~morrow/336_17/papers17/carson.pdf.

[2] D. Barden and C. Thomas. *An Introduction to Differential Manifolds.* Imperial College Press, London, 2003.

[3] M. P. Do Carmo. *Differential Geometry of Curves and Surfaces.* Prentice-Hall International Inc., USA, 1976.

[4] M. P. Do Carmo. *Riemannian Geometry. Mathematics: Theory and Applications.* Birkhäuser Boston, USA, 1992.

[5] EigenChris. Tensor calculus. https://www.youtube.com/@eigenchris.

[6] Ville Hirvonen. Christoffel symbols: A complete guide with examples. https://profoundphysics.com/christoffel-symbols-a-complete-guide-with-examples/.

[7] J. Lee. *Riemannian Manifolds, An Introduction to Curvature.* Graduate Text in Mathematics. Springer-Verlag, People's Republic of China, 1997.

[8] J. Lee. *Introduction to Smooth Manifolds.* Graduate Text in Mathematics. Springer-Verlag, USA, 2002.

[9] L. Auslander and R. E. MacKenzie. *Introduction to Differential Manifolds.* Dover Publications Inc., USA, 1977.

[10] L. C. Loveridge. Physical and geometric interpretations of the Riemann tensor, Ricci tensor, and scalar curvature (2016). arXiv:gr-qc/0401099.

[11] C. W. Misner, K. S. Thorne, and J. A. Wheeler, *Gravitation*, USA, 1973.

[12] M. Nakahara. *Geometry, Topology and Physics.* Institute of Physics Publishing, Great Britain, 2003.

[13] P. M. Gadea and J. M. Masqué. *Analysis and Algebra on Differentiable Manifolds.* Springer, London, 2009.

[14] B. Schutz. *Geometrical Methods of Mathematical Physics.* Cambridge University Press, United Kingdom, 1999.

[15] F. M. Stein. *Introduction to Matrices and Determinants.* Wadsworth Publishing Co., Belmont, 1967.

[16] S. Sternberg. *Lectures on Differential Geometry.* AMS Chelsea Publishing, USA, 1984.

[17] B. Vandereycken, P.-A. Absil, and S. Vandewalle. Embedded geometry of the set of symmetric positive semidefinite matrices of fixed rank (2009), pp. 389–392. doi: 10.1109/SSP.2009.5278558.

[18] R. M. Wald. *General Relativity.* University of Chicago Press, USA, 1984.

[19] P. R. Wallace. *Mathematical Analysis of Physical Problems.* Dover, USA, 1984.

[20] W. R. Parzynski and P. W. Zipse. *Introduction to Mathematical Analysis.* McGraw Hill, Singapore, 1987.

[21] Y. Choquet-Bruhat and C. Dewitt-Morette. *Analysis, Manifolds and Physics*, Part 1 and Part 2. Elsevier Science, Netherlands, 2000.

Index

Printed in the United States
by Baker & Taylor Publisher Services